T0073400

THE FOURTH INDUSTRIAL REVOLUTION

www.royalcollins.com

THE FOURTH INDUSTRIAL REVOLUTION

How the Modern World is Being Reshaped by
AI and the Internet

KEVIN CHEN

Books Beyond Boundaries

ROYAL COLLINS

The Fourth Industrial Revolution:
How the Modern World is Being Reshaped by AI and the Internet

KEVIN CHEN
Translated by Daniel McRyan

First published in 2023 by Royal Collins Publishing Group Inc.
Groupe Publication Royal Collins Inc.
550-555 boul. René-Lévesque O Montréal (Québec) H2Z1B1 Canada

ISBN: 978-1-4878-0983-6

To find out more about our publications, please visit www.royalcollins.com.

Preface

Modern civilization began with the Industrial Revolution. It has fundamentally changed the human mode of production. Compared with agriculture, which is subject to relatively limited output, commercial development must be based on industry. Among the three fields of industry, agriculture, and commerce, it is industry that truly plays a significant role in the continued prosperity of the economy and social stability. In fact, it is precisely the rapid development of industrial productivity in capitalist societies that has enabled the bourgeoisie to have created more productive forces in their less than one hundred years of rule than all the productive forces created ever before.

Industrial development has given mankind a greater ability to transform nature and obtain resources. People consume the products it produces either directly or indirectly, which greatly improves their living standards. It can be said that since the First Industrial Revolution, industry has governed the survival and development of mankind to a certain degree.

From the development of industrial civilization, after human society has gone through the three Industrial Revolutions respectively, in the age of steam, the age of electricity, and the age of information, digital technologies represented by the Internet and artificial intelligence are forming huge industrial capabilities and markets at an extremely fast speed, thus elevating the entire industrial production system to a new level and pushing human society into the Fourth Industrial Revolution.

Unlike the previous three Industrial Revolutions, the fourth one is an all-around innovation, a deep integration of cyber-physical systems, and a comprehensive reform of manufacturing technology and manufacturing mode. Moreover, during the Fourth Industrial Revolution, groups of new manufacturing technologies and manufacturing modes that follow the main thread of production method transformation keep springing up and integrating with each other, thus pushing human society into the pan-industrial era.

The Pan-industrial Revolution is a revolution that does not rely on a single discipline or a few types of technology, but on all-round multi-discipline, multi-technical level, wide-field synergy, and deep integration. It will widely extend to all industries, whether it is the consumer Internet or aerospace, daily life, or life science. It will be a multi-dimensional revolution embedded in the entire social system of technological economy as well as a revolution in advanced manufacturing modes that revolves around the application of advanced manufacturing technologies and radical changes in business practices. It will transform manufacturing concepts, manufacturing strategies, manufacturing technologies, and manufacturing organization and management in various fields.

The Pan-industrial Revolution will go beyond the Fourth Industrial Revolution and lay a new, upgraded path in the industrial world. Certainly, the opening of a new path brings new opportunities as well as challenges. Since the 1970s, there has been a wave of "de-industrialization" in the developed countries across the capitalist world. Industry makes prosperity, and recession too. Although in Western countries it was once regarded a wise move, now "de-industrialization" has done more harm. Therefore, "re-industrialization" is perhaps inevitable.

In the face of "re-industrialization," from the perspective of the countries, major industrial countries in the world have formulated corresponding strategic measures in recent years: Germany launched the reference architecture model Industry 4.0 (RAMI4.0), the United States released the Industrial Internet reference architecture (IIRA), Japan created the industrial value chain reference architecture (IVRA), and in China, the China Industrial Internet Industry Alliance published the "Industrial Internet System Architecture (Version 1.0)" in 2016.

From the perspective of the enterprises, a "lighthouse" is a new role in the pan-industrial age, and a model of "digital manufacturing" and "globalization 4.0." The emergence of "lighthouses" highlights the globalization of pan-industrial manufacturing in the future. A German company may set up a factory in China while a Chinese company may do that in the United States. Innovation does not distinguish between regions and backgrounds. From the procurement of basic materials to the processing industry, and to high-end manufacturers who cater to special needs, the industry varies greatly. It also means that companies of all sizes have the potential to innovate and stand out in the pan-industrial wave, whether it is a global blue-chip company or a local one with less than 100 employees.

The process of the Pan-industrial Revolution, whether it becomes intelligent, agile, informatization, and flexible, is no simple "technical substitution," but an organizational mode that makes the assembly line work that has been extremely refined or even differentiated since the Industrial Revolution "people-oriented" again. This means that the future production in the pan-industrial age will inevitably be human-centered. It requires the society to cultivate compound talents, the enterprises to adjust organizational structures, and the individuals to improve personal innovation capabilities.

The process of the Pan-industrial Revolution will also reconstruct the global value chain. Its reconstruction is an international division of labor that both upgrades and governs the value. In fact, with the profound changes in the international situation, the reconstruction of the global value chain is bound to happen. On the other hand, the rapid technological advancement catalyzed by COVID-19 in 2020 has added technological elements to it.

The Pan-industrial Revolution has promised an unprecedented technological future. This book aims to describe and explain this future pan-industrial world. Taking the Fourth Industrial Revolution as the background and the previous Industrial Revolutions as clues, it introduces the evolution of current manufacturing technology and manufacturing modes, deepens the understanding of the Pan-industrial Revolution by including "lighthouse" examples, and makes forward-looking predictions. It is both interesting and scientific, thus not obscure to read. While providing more information, this book aims to help readers to better understand the industrial age people were and are in, so that they can look forward to the future. As the industrial society reconstructs and the way of thinking is reset, the world will also change and have new meanings.

Contents

PART 2   THE PRESENT

PART 3   THE FUTURE

CONTENTS

PART 1
REVOLUTION

CHAPTER 1

No Industry, No Strength

1.1 The Rise of Industry

Without agriculture, there is no stability; without industry, there is no strength; without commerce, there is no wealth.

At the beginning of human civilization, its development relied on the materials that can be directly obtained in nature and used for consumption, such as plants and animals. Primitive humans used to dwell in either natural shelters or natural places that could be modified into shelters, like caves. From that, agricultural civilization came into being.

As agricultural civilization advances, mankind has gradually learned to process natural objects that could not be used for consumption directly. As they were made consumable, processing and manufacturing industries and construction industries were developed. Since then, an "industrial" society was born.

Certainly, in the early development of industrial society, mankind's transformation of matter was extremely rudimentary. It started with the transformation of low-level and single material's geometric shapes, such as grinding stones into sharp or blunt stone axes. Cavemen used it to attack beasts. They also sharpened sticks or utilized plant roots as "omnipotent" tools.

In the Mesolithic Age, stone tools have evolved into inlay equipment. For example, a wooden or bone handle was mounted on a stone axe, marking the transformation of a single material form into a composite form of two different qualitative materials. On top of that, composite tools such as stone knives, stone spears, and stone chains were developed, until bows and arrows were invented. In the Neolithic Age, humans learned to drill holes in stone tools, thus inventing stone sickles, stone shovels, stone hoes, and stone mortars for food processing.

At the same time, humans started to use tools for energy conversion. An obvious example is a hominid's understanding of the relationship between "fire" and themselves. From the fear of wildfires in forests or on grasslands caused by lightning strikes to learning to use fire to grill prey, and to using it to keep out the cold, illuminate, and keep away wild beasts, the mastery of artificial fire-making marks fire as a natural force that was truly utilized by human beings. When that happened, the development of the human body and the brain were promoted. As Engels pointed out, friction-making fire for the first time allowed humans to control a natural force, which ultimately distinguished them from the animals.

The use of fire further enabled hominids to make pottery. Pottery-making technique is a great leap of ancient material technology and material processing technology. For the first time, human beings could process materials beyond merely changing the geometric shape of materials, but commenced to change their physical and chemical properties. In addition, the development of ceramics laid the foundation for the upcoming birth of metallurgical technology.

As a result, from manual manufacturing to machine manufacturing, human society has witnessed increasingly complex economic activities that produce the "means of production," including "instruments of labor" and "objects of labor." In this way, industry has continuously developed into a huge "circuitous" production system for final direct consumption and use, a large part of industrial production activities are indirect and circuitous, and they produce for "instruments of labor" and "objects of labor." The "circuitous nature" of industrial production is in fact a high degree of division of labor in the production process. It is not only the division of various technologies, but also the general division of labor in society, so that it constitutes an intricate input-output relationship.

However, no matter how complicated the circuitous process of industrial production is, the nature of industry is to transform useless substances into useful and beneficial substances. The more developed the industry, the more material made into resources. In a highly developed industrial system, all substances can be turned into resources. Therefore, industry ultimately creates all the so-called "resources."

British sociologist Anthony Giddens pointed out in *The Consequences of Modernity*, published in 1990, that the modern industry constructed by the alliance of science and technology is changing nature in ways unimaginable for past generations; in the industrialized regions and gradually elsewhere in the world, human beings started to live in a humanized environment, which is certainly a materiality environment too, but no longer solely natural; not only the urban areas built, but most other areas have also become objects of human adjustment or control.

Evidently, industry has strong creativity, and it has penetrated into almost all walks of life, "industrializing" all fields of mankind's modern life. From agriculture, forestry, animal husbandry, fishery, transportation, to information transmission, culture, and art; from education, medical care, sports, fitness to leisure and tourism, and even military projects, industrialism is everywhere, everything relies on industrial technologies.

Industry's greatest contribution to mankind is that it is the carrier and necessary tool to achieve technological innovation. The greatest scientific discoveries and technological inventions of mankind, as well as the realization of outstanding human imagination, take industry as the basis and means. Technological advancement is the soul of industry, and industry is the body of technological advancement. Most technological innovations are manifested in industrial development or must be based on it.

Compared with agriculture, which is subject to relatively limited output, commercial development must be based on industry. Among the three fields of industry, agriculture, and commerce, it is industry that truly plays a significant role in the continued prosperity of the economy and social stability. In fact, it is precisely the rapid development of industrial productivity in a capitalist society that has enabled the bourgeoisie to have created more productive forces in their less than one hundred years of rule than all the productive forces created ever before.

Industrial development has given mankind a greater ability to transform nature and obtain resources. People consume the products it produces either directly or indirectly, which greatly improves their living standards. Therefore, only an industrial country can become an innovative country, and only by having a developed industry, especially a developed manufacturing industry, can a country lead in technological innovation. A scientific and technological revolution has the same destiny as the Industrial Revolution. So far, the age of industrialization marked by scientific rationality and technological advancement is the most glorious time of human development. It can be said that since the first Industrial Revolution, industry has governed the survival and development of mankind to a certain degree.

1.2 The Power of Revolution

1.2.1 The First Industrial Revolution: the Age of Steam

After the mid-18th century, the First Industrial Revolution broke out in the UK. It became the first country to industrialize itself.

In 1733, mechanic John Kay first invented the flying shuttle, which doubled the efficiency of weaving. Later, there was a contradiction between weaving and spinning, which caused a long-term "yarn shortage."

Therefore, in 1764, weaver and carpenter James Hargreaves invented the spinning jenny. It increased the spinning efficiency by 15 times and initially solved the contradiction between weaving and spinning. But the spinning jenny had a shortcoming— because it was spun by workers, the yarn came out thin and broke easily.

To overcome this shortcoming, in 1769, Richard Arkwright, a hairdresser and watchmaker, created a spinning machine that ran on water power, which changed the situation of workers spinning the machine. In addition, the yarn the water frame spun was resilient and thick, and could be used as a warp.

Because the water frame needed water power, factories had to be built by the river. In 1771, the establishment of the first cotton yarn factory broke through the old manual workshops, pushing the Industrial Revolution into the stage of modern machine factories.

To solve the problem of yarn being low count, in 1779, young worker Samuel Crompton, combining the advantages of the spinning jenny and the water frame, invented the spinning mule. It could spin 300–400 spindles at the same time, which greatly improved efficiency. In addition, the yarn it spun was both fine and robust.

As inventions continued to be made and the spinning machines to be improved, there was a surplus of cotton yarn, which in turn led to the invention of the loom. In 1785, engineer Edmund Cartwright created a hydrodynamic loom, which increased work efficiency by 40 times.

In 1791, the UK built the first weaving factory. With the invention, improvement, and use of cotton textile machinery, relevant processes have been continuously innovated and mechanized. For example, the cotton cleaning machine, carding machine, bleaching machine, and dying machine, were invented and widely used, one after another. Consequently, the entire system of the cotton textile industry achieved mechanization.

Moreover, with the rise of the Industrial Revolution, the textile industry made higher requirements for the power system. The old power (human power, animal power, natural power, etc.) could no longer adapt to the new production situation, so the invention of the steam engine became necessary and inevitable. As Marx put it, it was precisely the creation of machine tools that necessitated the revolution of the steam engine. In 1769, James Watt, an instrument maker at the University of Glasgow in Scotland, summed up the experience of his predecessors. After many trials and errors, he made the first single-acting steam engine. In 1782, he made an improvement on it and created the linkage steam engine.

The invention of the steam engine therefore became a significant symbol of mankind's first Industrial Revolution. It led mankind into the age of steam when there was modern machine production from the age of manual labor for the preceding two million years. The invention and application of the steam engine promoted the mechanization of various industrial sectors in the UK. The steam engine that was once only used for pumping water from mines was improved and used in the textile industry. In 1784, the first steam spinning mill was built in the UK. Afterwards, steam engines were used in the metallurgical industry, railway transportation, steam ships, etc.

The Industrial Revolution in Great Britain also enabled British social productive forces to take a giant leap. In the 80 years or so of the Industrial Revolution, the UK built a strong textile industry, metallurgical industry, coal industry, machinery industry, and transportation industry.

Mass production by machines has unprecedentedly enhanced labor productivity. From 1770 to 1840, the daily productivity of each worker in the UK went up on average by 20 times. From 1764 to 1841, the annual consumption of cotton in the UK increased from four million pounds to nearly 500 million pounds, up by over 120 times. From 1785 to 1850, the output of British cotton goods increased from 40 million yards to two billion yards, up by 50 times. From 1700 to 1850, coal production increased from 2.6 million tons to 49 million tons, an 18 times increase. From 1740–1850, the output of pig iron soared from 17,000 tons to 2.25 million tons, an 156 times increase. From 1825 to 1848, the length of the railway extended from 16 miles to 4,646 miles, up by 290 times.

At that time, the UK had not only basically broken away from the shackles of traditional handicrafts in the textile manufacture and mechanized it, but also replaced human labor with machines in fields such as transportation and metallurgy. By 1850 the UK accounted for 39% of the world's total industrial output value, and took up 21% of the world's total value of trade.

In merely a few decades, the Industrial Revolution enabled the UK to make a giant leap from a backward agricultural country to the world's most advanced capitalist industrial power, and earned it the title "world's factory." Marx and Engels pointed out that the bourgeoisie created more productive forces in their less than one hundred years of rule than all the productive forces created ever before.

It was the industrial foundation laid by the Industrial Revolution that made the Red Coats invincible all over the world in the 19th century, and these material guarantees were the cornerstone of the UK's building an empire on which the sun never set.

In addition to a major transformation in production technology and greatly enhanced productivity, the Industrial Revolution also changed social structure and production relations. Therefore, it was also a revolution of a social production mode. If the political revolution of the bourgeoisie only seized power from the feudal landlords, the fruit of the Industrial Revolution was the ultimate establishment of the capitalist system. Without the Industrial Revolution that transformed the society and economy, the capitalist system would have had no foundation nor would the political rule of the bourgeoisie be consolidated.

From the 1760s to the 1840s, the 80 years of the Industrial Revolution in Great Britain was a glorious feat in the history of the industrial civilization development, and the revolutionary power thereof has always been of great significance to this day.

1.2.2   The Second Industrial Revolution: the Age of Electricity

If the First Industrial Revolution in Great Britain is considered what made a powerful empire and earned the country the title "world factory" on the arena of the modern world economy, the Second Industrial Revolution, which began in the 1870s, with the invention and application of electric power, cast an epoch-making impact on the development of human society, triggering a worldwide Industrial Revolution and fundamentally changing production and lifestyles.

Primarily, the Second Industrial Revolution is marked by the wide use of electricity in production and life. In 1831, British scientist Faraday discovered electromagnetic induction, which was the theoretical basis of the invention of electricity. In 1866, Werner von Siemens made a modern dynamo. In 1870, Zénobe Gramme invented the Gramme machine as motor. In 1879, Thomas Edison switched on the first electric light that could be widely used.

The wide application of electrical inventions and new energy directly pushed heavy industry to take a giant leap, enabling large factories to obtain sustainable and effective power supply at great convenience but low cost. It made mass industrial production possible and laid the foundation for the subsequent economic monopoly.

Secondly, in the Second Industrial Revolution, the creation and use of internal combustion engines facilitated communication around the world. In 1885, Gottlieb Daimler and Karl Benz independently made the first automobile powered by an internal combustion engine. Thereafter, diesel locomotives, ocean-going ships, and aircraft have achieved rapid development. The invention of the internal combustion engine not only solved the power problem of transportation, but also promoted the rapid development of the chemical industry by addressing its insufficient long-term power.

Finally, after the creation of the wire telegraph, Alexander Graham Bell invented the telephone in 1876. In 1899, Guglielmo Marconi successfully sent a report between the UK and France. Consequently, the economic, political, and cultural ties around the world were further strengthened. Since then, communication tools advanced rapidly, enabling interpersonal communication to break through the limitations of distance and facilitating information exchange and transmission around the world. Thus, economic, political, and cultural ties around the world have been stronger.

The huge productivity created by the Second Industrial Revolution pushed capitalism from a free stage to a monopoly stage, accelerated the process of economic globalization, and formed the world market and world economic system. However, it also triggered violent imperialist aggression, made international politics from regional to global, and formed the capitalist world system.

Specifically, the Second Industrial Revolution led to changes and re-ranking in the strength of the great powers. By the beginning of the 20th century, the strength of the United States and Germany were ranked at the top because of their improvement and application of the technologies in the Second Industrial Revolution that were created at the end of the 19th century.

The truth is that rarely were the many inventions in the First Industrial Revolution, such as those in the textile industry, mining industry, metallurgical industry, and transportation industry made by scientists. Instead, it was technicians who made most of them. The First Industrial Revolution adopted the practical experience of the technicians to make up for the missing parts in theories. There was no full integration of science and technology. But the situation after 1870 changed drastically. The new development of natural science started to be closely integrated with industrial production. Science began to play an increasingly important role in invention, becoming an important part of mass industrial production.

The specific technical characteristics and geographic nature of the Second Industrial Revolution led to the gradual collapse of the old European order. Therefore, contrary to Spain, Portugal, and Holland, which declined to be second-tier countries, the United States, France, Germany, Austro-Hungary, Italy, and Russia, which achieved industrialization in the Second Industrial Revolution, stood out in international politics.

As major capitalist countries such as those in Europe, the United States, and Japan entered the stage of monopoly capitalism, imperialism began to carve up the world according to capital and power. The newly rising countries Germany, the United States, and Japan demanded a change in the status quo of international politics, while the old imperialism strove to maintain vested interests. This political confrontation inevitably ignited conflicts and wars among the great powers.

As a result, the UK gave up its long-abiding "splendid isolation" policy and made adjustments to its world policy. The European powers launched a fierce arms race, which greatly accelerated the reorganization of strength between the world powers. At the beginning of the 20th century, a two-tiered pattern of confrontation between the Anglo-French-Russian Allied Powers and the German-Austrian Allied Powers broke out in Europe. The intensification and escalation of conflicts eventually led to the outbreak of World War I. The war weakened European powers badly, and the matchless European hegemony came to an end. With the rise of new countries and the decline of traditional powers, the old European international system moved out of Europe and expanded globally.

In contrast to the First Industrial Revolution that made the UK an "empire on which the sun never sets," the second injected a strong and sustained impetus into the formation of a global international political system. Driven by the Industrial Revolution, new powers such as Germany, the United States, Japan, and Russia have stepped onto the international political arena and took center stage in the international system. The Second Industrial Revolution also disintegrated regional orders such as the old European international order, the Muslim world, and the East Asian tributary system.

In general, the Second Industrial Revolution, with its extraordinary scientific and technological achievements, has fueled the development of the modern world and fundamentally changed its pattern. When the advancement of science and technology promoted economic development, it was also quietly changing the bargaining chip of the big powers in the game, and paving the way for changes in the global landscape.

1.2.3   The Third Industrial Revolution: the Age of Information

Half a century ago, the Third Industrial Revolution broke out. Based on the development of intelligentization, digitization, and information technology, it mainly transformed large-scale assembly lines and flexible manufacturing systems with modern basic manufacturing technology. Characterized by personalized manufacturing and rapid market response that are built on reconfigurable production systems, the Third Industrial Revolution is a profound reform in the technological and economic paradigm embedded in technology, management, and institutional systems.

The occurrence and development of the Third Industrial Revolution resulted from the joint drive and collaboration of exogenous technological progress and endogenous national policies.

Judging from exogenous factors, the Third Industrial Revolution is first of all the inevitable outcome of exogenous technological accumulation and technological innovation entering a specific cycle and stage. The rapid advancement of the lowest level technology in the modern manufacturing technology system—information technology has lowered the cost of information storage, transmission, and processing geometrically. From 1992 to 2010, the average transmission cost of 1 MB of data plummeted from $222 to $0.13; the cost to store 1 GB of data dropped sharply from $569 to $0.06.

Information's better industrial service capacity and lower use cost have massively promoted the maturity of basic manufacturing technologies based on information and communication technology (ICT) such as artificial intelligence, digital manufacturing, and industrial robots. And their maturity and lower costs have further widened the application of these cutting-edge manufacturing technologies in large-scale assembly lines and flexible manufacturing systems, and inspired new manufacturing systems, such as reconfigurable production systems and 3D printing, to be created by integrating peripheral technologies such as new materials, new energy, and optoelectronics. This multi-level and multi-field technological innovation and interaction together constitute the basic skeleton of the technological evolution of the Third Industrial Revolution.

The fundamental technological driving force that deepens the Third Industrial Revolution lies in the innovation and breakthrough of basic manufacturing technologies such as digital manufacturing, artificial intelligence, industrial robots, and additive manufacturing. The group breakthroughs and application conditions of a series of relevant key technologies such as rapid prototyping technology, new material technology, industrial robot technology, and artificial intelligence technology have gradually matured, thus effectively propelling a significant leap in manufacturing efficiency and taking the entire production system to a whole new level.

In addition to the exogenous drive from technology, the continuous deepening of the Third Industrial Revolution is also the fruit of endogenous institutional arrangements and policy designs that reflect the strategic intentions of major industrialized countries.

During the financial crisis, the stable performance of the German economy, which has always emphasized the development of the real economy, and the speedy recovery of the Chinese economy, which has the world's fastest growing manufacturing industry, are in sharp contrast with a sagging economy and even the debt crisis of most western countries. This reality has convinced major industrialized countries to reflect on the economic functions and strategic significance of their manufacturing and manufacturing industry in their national innovation system and industrial system.

From but Beyond Technology

The Third Industrial Revolution happened because of breakthroughs in manufacturing technologies, but its mechanism and effects on industrial economy are never confined to merely manufacturing technologies. The replacement and innovation of traditional manufacturing modes by advanced manufacturing technologies, manufacturing systems, and manufacturing paradigms under the Third Industrial Revolution will fundamentally change the function and nature of the core "productive assets" of industrial enterprises—manufacturing not only determines the production cost, but also affects a company's product innovation capabilities and dynamic efficiency directly. The importance of knowledge in the manufacturing system, compared to equipment and general labor, is obviously greater.

On the one hand, the Third Industrial Revolution shifted enterprises from mass production to mass customization.

Single-piece manufacturing in small batches was the starting point for the development of industrial production. It was characterized by production being carried out totally in accordance with the individual requirements of different customers. Technicians could only make, using general machinery, one or a few non-standardized products at each production. The application range and technical complexity of such a manufacturing paradigm peaked at the end of the 19th century.

Back then, a considerable number of horse-drawn carriage manufacturers in Europe and the United States switched to automobile production. Since the production of components was highly dependent on the personal skills of technicians, the production of auto components, car body building, and assembly were largely scattered in workshops equipped with general-purpose machines. Under this manufacturing paradigm, the basic business model of the manufacturer is: first present the automobile design concepts to the customer, and the customer selects one and signs an order with the manufacturer; then conduct detailed product design according to the design concept and customer requirements; and finally produce and deliver products according to product design. As each link of sales, design, and production was highly personalized, the output of manual production was quite limited.

Mass production is one of the most important accelerators to promote the development of the industrial society. It is the fruit of the Second Industrial Revolution. The core content thereof is to use assembly lines composed of specialized equipment to mass-produce standardized products. The specialization and standardization of mass production not only significantly lowers the production cost, but also greatly improves the precision of the product.

The strong economic vitality of mass production is that by reducing production costs, it expands market demand, which in turn makes more space for mass production, thus forming a mechanism of mutual enhancement between market demand and production scale. The characteristic of mass production is summarized as "great quantity but few varieties."

Mass customization is a production mode transformation inspired by the integration of information technology and manufacturing technology in the 1980s. As an outcome of the inoculation stage of the Third Industrial Revolution, it involved a substantial increase in the variety of products to meet the broader personalized needs of consumers, which allowed user innovation and creativity to play a more significant role in the development of the industry. Suppliers' domination of industrial innovation was weakened, and the competitive strategy of companies relying on economies of scale to reduce costs was challenged.

Moreover, as mass customization emphasizes the diversity of products, the efficiency and flexibility of the entire supply chain have become the key to the competitiveness of products and enterprises. A vertical organizational structure of the industry, rather than a horizontal one that takes market concentration as the main measurement under the mass production paradigm, has become the major determinant to the overall efficiency and competitiveness of the industry.

On the other hand, the Third Industrial Revolution drove companies to shift from a rigid manufacturing system to a reconfigurable manufacturing system. The traditional rigid manufacturing system was composed of dedicated automatic production equipment. It was designed to have a fixed configuration after the operation, thus more suitable for the production of a single product. The flexible manufacturing system matches with the production of various products in small quantities. The entire system is expensive, thus the high production cost. Due to the incompatibility between the software operated by different equipment manufacturers, it is difficult to integrate and operate the system.

In the Third Industrial Revolution, new manufacturing systems represented by the reconfigurable manufacturing system were adapted to mass customized production. This type of manufacturing system achieves fast debugging and producing by rearranging, reusing, and updating system configurations or subsystems. It has strong compatibility, flexibility, and outstanding production capacity.

Reshape the International Landscape

The history of industrial technology development shows that the advent of new manufacturing paradigms that are compatible with new technologies and economic conditions are accompanied by both the advancement of manufacturing technology and new human capital investment. It has adjusted corporate strategic directions and investment structures, and encouraged new forms of industrial organization to spring up.

Therefore, the Third Industrial Revolution is also a comprehensive coordinated transformation of technology, management, systems, and policies in the sense of a techno-economic paradigm. This transformation will enormously adjust the industrial organization structure, industrial competition paradigm, and global industrial competition pattern in the end.

The Third Industrial Revolution is reshaping the international industrial division of labor. Second-mover countries must seek new industrial paths to catch up. The process of wide application of modern manufacturing technology and production equipment in the Third Industrial Revolution is a process where "modern machinery and well-educated employees" replace "traditional machinery and simple labor" step by step. The economic rationality of this replacement lies in not only that modern manufacturing has improved the marginal productivity of labor, but also that the products modern manufacturing systems produce have better performance, more powerful functions, and shorter development cycles.

Modern manufacturing has reduced the industry's reliance on simple labor while giving products richer competitive elements – the value creation ability of manufacturing, so that the strategic position thereof in the industrial value chain becomes as important as R&D and marketing, and weighs more than other values creation links.

Therefore, developed industrial countries can control the new industrial commanding heights by developing modern equipment such as industrial robots, high-end CNC machine tools, and flexible manufacturing systems, improve the production efficiency of traditional industries by equipping them with modern manufacturing technologies and manufacturing systems, and strengthen the engineering and industrialization capabilities of new technologies by equipping emerging industries. Simultaneously, the deep integration of modern manufacturing systems and service industries (such as open software communities and industrial design communities) may further strengthen the leading advantages of developed countries in the high-end service industry.

The Third Industrial Revolution is a multi-dimensional transformation embedded in the technological, economic and social systems. It has had a profound impact on international relations: on the one hand, it has aggravated the imbalance of development of capitalist countries, changing their international standing and leading to the strong rise of the United States as the only superpower; on the other hand, it has given socialist countries a powerful driving force in the struggle against western capitalist countries.

Apparently, the first three Industrial Revolutions have had an enormous impact on world politics, economics, science and technology, military, etc. They have fundamentally reshaped the international landscape. While each Industrial Revolution brought about major changes in the global economy and society, changes took place in the national strength and competitive position of countries around the world. Some have risen because of it and become the dominant players in certain fields and even the world economy and global governance, while others have missed the opportunities, lost prosperity, and declined. History marches forward in such a cycle and presents answers in it.

1.3 Upheaval Has Taken Place, Upgrading the Industry

The Industrial Revolution, as the starting point of modern civilization, has fundamentally transformed the production mode of mankind. Under the long-term accumulative effect of the previous three Industrial Revolutions, the Fourth Industrial Revolution, which is characterized by intelligence and represented by cutting-edge technologies such as artificial intelligence, quantum communication, biotechnology, and virtual reality, is rising faster than ever. The curtain of a new round of global technological competition is being lifted.

The Fourth Industrial Revolution is another major event that has greatly changed the economic life of human society following the example of the age of steam, the age of electricity, and the age of information. This technological revolution revolves around the in-depth integration of networking, informatization, and intelligentization. While enhancing productivity and enriching material supplies, it will also reshape the labor forms and requirements that combine human labor and machine power, and add new content and new methods to industrial policies.

First of all, in the Fourth Industrial Revolution, the advancing speed of various cutting-edge technologies is exponential. In the past, the upgrading of technologies might take a few years or over a decade, but now, a round of technological innovation is complete within a couple of years.

According to the authoritative American futurist Ray Kurzweil, tens of thousands of years ago, technological growth was so slow that a generation of people could not see obvious results; during the recent 100 years, one person could at least witness one huge technological leap; and since the dawn of the 21st century, every three to five years, there are changes similar to the sum of the technological achievements in the history of mankind. The speed of technological advancement has gone beyond imagination.

Secondly, plenty of new technological achievements have entered people's daily production and lives. They have had a profound impact on their thinking, culture, life, and external exchanges, and next deeply affected politics, economics, technology, diplomacy, and society. For example, the development of artificial intelligence, from auto-pilot cars to drones, and from virtual assistants to automatic translation, has penetrated almost every aspect of people's lives.

The wide application of endless new technologies to the military domain has built a strong national defense. Their application to the economic sphere has been constantly inspiring new economies, new products, new rules, and new business forms. The traditional mode of production and life has been completely changed, and new political and economic systems are being reshaped.

Finally, the value-added sector of the industry has expanded from manufacturing to service. With technologies such as big data and cloud computing, the key to the competitiveness of industrial enterprises and the main source of profits will be the capability of data analysis, software, and system integration. Using big data to study customer or user information can open up new markets for enterprises, thus create bigger value.

For example, General Electric used to be a manufacturing company, but it has expanded its business to technology, management, maintenance, and other services, whose output value has actually exceeded two-thirds of its total output value. Clearly, with the help of big data, equipment manufacturing enterprises offer to provide equipment user enterprises with predictive maintenance solutions and services, meaning they extend the service chain to achieve stronger competitiveness and greater value.

Apparently, unlike the previous three Industrial Revolutions, the fourth is an all-round innovation. Although it is common sense today that a modern industrial system consists of energy, transportation, and manufacturing, in the early days of the Second Industrial Revolution, all sectors of industry, academia, and governments, when faced with a variety of technologies, had no idea how to use them to profit, nor what kind of system should be integrated to accommodate them.

The most outstanding features of the Fourth Industrial Revolution are the integration of multiple technologies, the formation of a new industrial chain logic, and the technology-commerce fission effect. Together, they will eventually become the basis for cities, enterprises, and industries to enhance the general production relations. For example, the re-conceived services and business models supported by smart manufacturing enable enterprises to simplify the production relationships with suppliers, manufacturers, and customers throughout the value chain. It also adopts technology to unify people, processes, and products. Manufacturers and service organizations can access massive data like never before, thus easier to understand, control, and improve all aspects of their operations.

As the Fourth Industrial Revolution marches forward, history will be at a new turning point. The Fourth Industrial Revolution is all-round innovation. Should countries aspire to lead the new round of Industrial Revolution, they must overcome obstacles, accelerate integration, and develop comprehensively. The curtain of the fourth technology competition is being lifted.

CHAPTER 2

Into the Pan-industrial Age

2.1 Time for the Industrial Platforms

The platform economy is both a new way of organizing productivity in the digital age and a new driving force for economic development. The platform is where the accumulation, distribution, circulation, and cross integration of production factors take place. By using information tools, instant messaging, and network functions, it can connect goods, service providers and customers anywhere in the world. Today, different kinds of platforms have reformed one consumer market after another. As the Fourth Industrial Revolution deepens, it is time for the industrial platforms.

2.1.1 From Consumer Platform to Industrial Platform

The development of information technology has digitalized the physical world of human society. In the 1990s, the digital revolution was ascendant, starting the first wave of the digital economy. During that, digital technology was applied commercially in the consumption area on a large scale. Almost all end users of major business models such as portals, online videos, online music, and e-commerce are consumers. This stage is therefore called the "consumer-oriented Internet." And the consumer platform is an important foundation for the consumer-oriented Internet era.

The structure of the traditional industry presents a "V" shape. One side is the supply, the other the demand, and the middle the goods or services that act as an intermediary between them. Many suppliers and a considerable number of consumers are separated on both sides. Due to the difficulty in identifying each other, coupled with information asymmetry, inefficient transactions and waste of resources happen. The consumer-oriented Internet has obtained the advantage of resource integration with the help of the consumer platform, thus greatly improving the communication efficiency between the supply and demand sides. This advantage is mainly reflected at the two ends of "V." The link between the suppliers and the consumers is the commodity, and it breaks off once the transaction is completed.

Benefiting from the characteristics of the platform, consumer platform-based enterprises possess energy that traditional enterprises do not, including non-competitiveness, network effects, scale effects, and economies of scope.

From the perspective of non-competitiveness, the operation of traditional enterprises mostly relies on consumable means of production, and the marginal cost of their economic activities cannot be reduced to zero, thus limiting the effect of scale. For consumer platform-based enterprises, data as a means of production almost costs nothing to duplicate and transmit. And the use of data by any single user does not affect that of other users, nor does it increase the supply cost of the enterprises. For example, for an online game, when the number of online users increases, the new cost is almost zero. Therefore, consumer platform-based enterprises can serve many users without a clear upper limit to their capacity and scale effects.

In addition, social platforms have a typical same-side network effect. Out of the need to socialize, users tend to prefer social platforms with more users, such as Facebook or WeChat. Because the platform precipitates social relationships, the switching cost for a single user is high, and sometimes the platform is irreplaceable. Bilateral markets such as e-commerce have a strong cross-edge network effect. The great number and various types of merchants can attract more consumers to shop on the platform, and the increase in the number of consumers will attract more businesses to join, thus achieving cross-edge, indirect network effects.

In terms of the scale effect, on the one hand, data has a scale effect. For example, Beke Real Estate: once the information of housing resources is uploaded on Beke, no matter how many users access it, there is almost no difference in the cost. On the other hand, technology also generates a scale effect. Take Alibaba Cloud as an example. Once the functions on its platform are developed, no matter how many customers use them, the marginal cost is low. Conversely, a larger scale can fund the platform more to support subsequent R&D and upgrades, thereby forming barriers to competition.

Also, because the consumer-oriented Internet does not consider how commodities are produced, the application scenarios are relatively simple, and the requirements for network performance are relatively less strict. Meanwhile, as complex and diverse production processes are not involved, the application threshold is low and homogeneous, the development model of the consumer-oriented Internet is highly reproducible, and the scale effect is easily and quickly achieved. In addition, the investment payback period of a consumer-oriented Internet is usually shorter, thus easier to obtain social capital support. This has also allowed the consumer-oriented Internet to gain rapid growth with the support of digital technology.

Finally, for economies of scope, Bilibili is the leading video platform for animations and games in China. By analyzing user viewing behavior on the platform, Bilibili can locate what animation is trending and what game types are well accepted, so that it can develop animation and gaming business and launch the most suitable content for distribution according to user needs on the platform in the future. Economies of scope enable platform-based enterprises to extend their business scopes, thus further promoting the birth of large-scale platform-based enterprises.

In fact, it is the Internet-based companies that lead and drive the development of the consumer-oriented Internet. It can quickly integrate suppliers and consumers according to the grade and category of goods. Essentially, this is the platform model. With the powerful network effect thereof, all users may obtain higher value in the process of network expansion. This is the first half of the Internet, the ecosystem of the consumer Internet.

For example, from B2C to C2C, depending on the Internet platform, Amazon, which started as an online bookstore, quickly grew to sell everything. It connects millions of consumers and thousands of manufacturers, distributors, and retailers of almost all types of products. With an enormous network and data analysis, it renders services spanning from a bookstore to a supermarket, including cloud computing, data storage, and an increasing number of physical sales.

In addition to Amazon, although Apple is still best known for its electronic products, in fact, it has also evolved from a simple communication tool provider to what connects thousands of data, entertainment and service providers owing to the empowerment of the platform. It includes everything from publishers and music companies to movie studios, game makers, and application designers.

In the first half of the Internet, the platform model demonstrates how powerful the network effect is, and that it is likely for all users to obtain higher value during network expansion. The reason why Facebook can attract hundreds of millions of users is precisely because users can find the greatest number of different people there. In turn, this generates huge advertising revenue to Facebook, as well as revenue from games, apps, and other products sold to members.

Therefore, in the age of the consumer-oriented Internet, Internet vendors that have established platforms have all gained superior competitiveness—the consumer-oriented Internet platform enables them to not only produce goods or directly render services to customers, but also connect service providers and consumers in need of goods and services. Once the platform reaches a certain critical number, it enters a virtuous circle, and the number of participants will continue to grow to create more value.

However, the consumer platform dividend is slowly fading. On the one hand, processing and manufacturing industries need urgent transformation as the "Internet +" trend deepens, industrial enterprises suffer from a continued downturn in the international market, domestic economic growth slows down, demographic dividends disappears, there are higher requirements for energy conservation and environmental protection, and customers make more stringent demands. On the other hand, under the pressure of the rapid changes in the market structure and more fierce differentiated competition, the iterative update of technology, the rapid upgrade of market demand, and the active innovation of business models are forcing industrial enterprises to seek a new way out.

Under these circumstances, the industrial platform, as the foundation for the digital, networked, and intelligent development of the manufacturing industry, once again set off a new wave of exploring the integration of information technology and manufacturing, leading the platform economy to transit from a consumer platform to an industrial platform.

1.2.2 Industrial Platform: More Complex, More Diverse, and More Powerful

As the platform and flexibility of manufacturing in the Fourth Industrial Revolution, to meet various customized production, the industrial platform is an important carrier for developing the core intelligent manufacturing technology of intelligent manufacturing.

The industrial platform differs greatly from the well-known consumer platform of today with its excellence in increasing manufacturing speed, accuracy, efficiency, and flexibility. Its structure and function are more complex. And it will operate in ecosystems and markets that are totally different from the consumer domain that the most successful platforms have occupied so far.

More Intricate Business Ecosystems

The industrial platform will serve a large and intricate business ecosystem. As a market participant, it, like other market entities, is first of all an enterprise, which is a legal entity that aims to profit, that makes use of various production factors to sell goods or services in the market, that operates independently, assumes sole responsibility for its profits and losses, and performs an independent accounting. However, unlike other market entities, the goods or service that platform enterprises sell in the market is organizing the market itself, that is, organizing interactions and matching for bilateral or multilateral groups. Therefore, the platform plays the role of market organizer.

It is the fact that the platform has the dual identities of market participants and organizers that convinces many economists that Coase's classic theorem on the nature of the firms—the blurred boundary between the enterprise and the market as the two ways of allocating resources. The platform is both an enterprise and a market. Therefore, some economists directly refer to the platform as "the concretization of the market." The market concept as a resource allocation method can be simply summarized as the mechanism to allocate a number of resources, such as price mechanism, transaction mechanism, and competition mechanism.

Platforms being the "concretization of the market" means that almost all platforms play an important role in two or more mechanisms, casting a huge impact on the allocation of resources. For example, the Taobao platform plays a role in the setting of transaction mechanisms, the maintenance of competition order, and the construction of a credit system, while the Didi platform participates in the construction of the four mechanisms of price, transaction, competition, and credit.

Therefore, the industrial platform will also have an influence on the allocation of market resources. It faces at least four types of users, including:

Direct platform users are the platform owner and many companies that use various elements of the platform, including manufacturers, supplier logistics companies, wholesalers, retailers, design companies, marketing consultants, and other service providers.

Indirect platform users are organizations that interact with the platform or with the direct users, including regulators, tax authorities, and other government departments, as well as university-affiliated research offices, and companies that render financial, legal, accounting, and other professional services.

The communication network is the internal system the platform owner offers to platform users, including Wi-Fi, short-range wireless communication, Bluetooth, wireless routers, unlimited range extenders and repeaters, and external communication networks, such as telecommunications companies, Internet providers, Internet service providers, content delivery networks and independent Internet of Things networks.

The end product users connected to the platform refer to the customers and user companies of the platform owner.

Compared with the consumer-oriented Internet, the business ecosystem of industrial platforms usually involves hundreds or thousands of organizations, and millions of individual participants in their management. This will be a more complex world.

Greater Diversity in User Interactions

If the consumer platform is considered what enables both supply and demand on both sides to identify and match each other remotely, the industrial platform needs to sink so that consumers can participate in close proximity, and to render personal services to the suppliers. For the suppliers, it needs to sink to every link of the industrial chain, pierce the boundaries of the enterprises, go deep into their daily operation, and highlight the status of the entity on the suppliers. On the demand side, consumers are involved in production, design, and innovation.

Specifically, the consumer platform serves clients. It changes people's lifestyles, and its target market is 830 million Chinese netizens and 1.4 billion Chinese people, while the industrial platform serves businesses. In a strict sense, its service objects are various organizations, including market entities such as enterprises, individual industrial and commercial units, and rural cooperatives, as well as governments, schools, hospitals, other institutions, and social organizations. It has changed the production, operation, and management of the society. In China, there are 120 million market entities alone.

This also shows that the industrial platform builds a long chain. From elements to value, to complete it, multiple ecological communities such as customers and service providers are needed. Service providers integrate elements of the industrial platform into solutions and offer them to traditional enterprises as customers. According to the solutions, traditional enterprises can promote the interconnection and ecologicalization of their internal operating processes and even the processes on the assembly line, and ultimately render consumers with personalized services.

Stronger Network Effect

Compared with the network effects of the consumer platforms, those of the industrial platforms are quite different and maybe more powerful. These network effects will prepare for the rise of pan-industrial companies.

As mentioned earlier, the all-around rise of the consumer platform economy is mainly due to the tremendous reduction in transaction costs thanks to the Internet-centric information technology, which maximizes the network effects from the platform model itself, especially the cross-edge network effects. The value of the network effect to users as a network product or service depends on the number of other users in the network. Emails and WeChat are typical examples. The further expansion of the consumer platform economy has generated cross-edge network effects, that is, the value of a platform product or service to users depends on the scale of users on the other side of the platform. For example, the more drivers there are on the online car-hailing platform, the greater the value of the platform to passengers. Similarly, the more WeChat users there are, the more attractive the WeChat official account or WeChat Moments ads will be to businesses. This cross-edge network effect is the core advantage of the platform model over the traditional non-platform business model.

In the age of industrial platforms, this cross-edge network effect will grow even stronger. Industrial platform owners will want to create a larger community on both the corporate side and the consumer side. This will allow them to enjoy more benefits from the network effects, including those of growing from the interaction between corporate networks and consumer networks. An enormous and growing corporate network means a wide range of information, goods, and services to attract more consumers; and an enormous and growing consumer network attracts more enterprises who want to sell goods to more consumers. The growth of either side's platform will help the other side's growth in other aspects.

When a new electronic product retailer joins an industrial platform, it will "take" all the company's consumer customers to it. They will be able to receive messages and purchase invitations from the platform, and become the sales targets of related products, services, components, and upgraded products. They will also allow the platform to obtain additional information on consumer preferences, shopping habits and browsing patterns, thus making it easier for its corporate users to issue purchase and sale invitations and effectively target new customers. Therefore, the self-enhancement of the network effect at both ends helps the platform grow bigger and stronger.

This also means that the managers of industrial platforms must have the skills to build and maintain large-scale networks of both enterprises and consumers. This kind of job that requires multiple skills is certainly more challenging than managing only a consumer platform, and also more rewarding.

In addition, business users will also be able to generate other valuable network effects on industrial platforms. Many of them are experienced product designers; some have engineering talent or practical technical knowledge; others are experts in marketing, sales, logistics, services, and other important business activities. Well-run industrial platforms will find ways to use these information and conceptual resources, and platform managers may initiate co-creation, collaboration, and outsourcing that can derive valuable ideas for other platform users.

Lastly, as the number of business users grows, the network effect of industrial platforms will also widen. Companies that purchase the same materials, such as the same metal powder used in 3D printing, can use this platform to gather their orders and receive volume discounts, special transportation and warehousing services, and other preferential business conditions.

Companies that produce related goods or render services for overlapping customer groups will use the influence of this platform to create attractive combo products—companies that make baby clothes, baby furniture, diapers, toys, and publish children's books can work together to develop neonatal products that can be sold on the consumer platform, or prenatal combo products.

Companies in the same market can also accumulate value by sharing consumer information. Data collected from shopping activities, browsing history, and other places on the Internet of Things can enable companies to conduct an in-depth analysis of existing and potential customers. The information obtained from it can help manufacturers make products that better cater to customer needs and market them more effectively.

There is almost no limitation to the network effects that industrial platforms can generate for business users. Due to the network effect, a platform with a higher market share can bring more value to users, attract and gather more users, and better keep the existing users stay on the platform. Most importantly, companies that own the best industrial platforms and enjoy the benefits of self-enhancement due to huge network effects will be in a favorable position and grow into giants in the future pan-industrial world.

2.2   Build an Industrial Platform

In different fields, what is called "platform" is not the same. At the product level, "platform" usually refers to the company's project to develop a new generation of products or a certain series of products. Wheelwright and Clark were the first to use "platform products" to describe new derivative products that can change or replace certain features of the original products while still being able to meet the needs of core customers. At the technical system level, "platform" is defined as a key point in the industry that has great value and controls the industry, such as the operating system in the computer industry and the kernel of the browsers. At the transaction level, economists use the term to characterize institutions or companies responsible for intermediary transactions between two or more parties.

Although it is many completely different things in different fields that have been named "platform," such as software programs, websites, operating systems, car bodies, and game consoles, they share some common characteristics. For example, most of the definitions of "platforms" emphasize reusable and shared elements in products, industries, or systems.

Meyer and Lehnerd believe that a platform is a set of reusable common components, on which companies can effectively create a series of derivatives, and reusable elements are only part of the platform's system architecture. Wheelwright and Clark pointed out that the platform system architecture includes a peripheral component that can install additional functions and continue to improve itself on the basis of core functions so that targeted derivatives are produced for market segments. Whitney et al. defined the platform system architecture more accurately, pointing out that it includes: a list of functions, peripheral components used to implement different functions, interfaces between different components, and a description list of the operating effects of the system under different conditions.

To sum up, the basic characteristic of platform system architecture is: some core components remain basically unchanged throughout the life cycle of the platform while other peripheral components are diversified. Meanwhile, necessary interfaces are required between different components, meaning that the common feature of platform system architecture is composed of three parts: stable core components, diverse peripheral components, and interfaces between components.

Among them, the core components are used repeatedly. As the market environment changes, there is no need to design or rebuild the system from scratch. Economies of scope can be created by increasing the diversity of peripheral components and simultaneously developing multiple derivatives for market segments; while the mass production and repeated use of the core components amortize the fixed cost of the entire product series or industrial evolution, they also realize economies of scale; finally, improving the standardization of the interface can reduce the cost of compatibility between peripheral components and core components, which further lowers the cost of the products. Apparently, the platform achieves its superiority via the product technology architecture expressed by modularity, connectivity, and interface standards.

The industrial platform born from platforms is a deep cross platform that involves intelligent manufacturing and the industrial Internet in the construction of new infrastructure. Within it, the core layer, basic and versatile, is a long-term unchanged module in the platform architecture, such as core intelligent manufacturing technologies in artificial intelligence, big data, cloud computing, etc.; the application layer is the diverse periphery module of the platform architecture, which aims to adapt to diverse and changing demand scenarios, such as derivative products in market segments, micro-service modules in industrial software systems, personalized expansion services for platforms, and industrial Apps customized for enterprise, etc.; the interface layer includes the data source and core layer, the data source and application layer, and the interfaces between the core layer and the application layer, and between the internal entities of each layer architecture, such as gateways, communication protocols, industry standards, data conversion, etc.

The industrial platform will no longer be limited to traditional CPUs. It performs big data analysis to advance technology research and development and cutting-edge product design, and builds smart enterprises that create smartly. It not only applies digital technology to update industrial manufacturing equipment, uses big data, the Internet of Things, and blockchain to upgrade new infrastructure such as application vendors, cloud service providers, IDC services, etc. but also uses industrial integrated data to predict future trends of the industry so that the development strategy is customized and the flexible production system further improved. It is the basic guarantee for the development and extension of the Fourth Industrial Revolution, and also the core link that connects enterprises and users in the age of big data.

To build an industrial platform, enterprises need to get rid of technological dependence, avoid excessive inertia in technological innovation, and stay alert to excessive motivation in the innovation of technological games. At the same time, it is more necessary for them to improve the efficiency of technological research and development on the basis of integrating production resources and user resources, to increase technological marginal income, and to promote the realization of the positive externality of technology spillover. As a result, there will be an endogenous innovation reform of the enterprise's industrial platform and a network paradigm for industrial innovation, which ultimately form the ecological paradigm of industrial regional innovation and cluster innovation.

2.2.1 Promote Industrial Productivity

Industrial platformization is not a social process that arises out of thin air, but what, with the support from previous technological accumulation, takes advanced technologies such as artificial intelligence and a new generation of information and communication technology as the turning point of industrial transformation.

To promote industrial productivity, new technologies are required first, such as artificial intelligence as technical support. The key to manufacturing intelligence is "intelligence," which relies on artificial intelligence. So, artificial intelligence is the supporting technology of manufacturing intelligence. Take discrete manufacturing as an example. It is characterized by scattered equipment and discontinuous processes. To promote the intelligentization of discrete manufacturing, it is necessary to collect basic data through smart sensors, and manufacture flexibly through smart machine tools, industrial robots, and smart warehouse systems, so as to make products intelligent and increase value. This series of processes are all based on artificial intelligence and achieved via intelligent transformation of each manufacturing link.

Second, the industrial Internet is used as a way of connection. Enterprise intelligentization does not mean industrial intelligentization. From enterprise intelligentization to industrial intelligentization, all enterprises must be connected. Therefore, the industrial Internet is what connects the intelligentization of the manufacturing industry. It establishes a connection between the machines, equipment, networks, and workers involved in manufacturing through the Internet, realizes the full interconnection of humans, machines, and objects, and quantifies manufacturing activities and links based on a variety of intelligent predictive algorithms. Consequently, it builds a huge industrial Internet, takes it as the link for the coordinated development of manufacturing enterprises, and continues to promote the intelligentization of the manufacturing industry.

Finally, a new manufacturing system is to be built as a development goal. The progress of manufacturing intelligentization not only requires manufacturers to intelligentize, but also depends on the development of other related industries. In other words, it is necessary to build a new manufacturing system as the developmental goal of industrial intelligentization. With the intelligentization of manufacturers as the main development direction, related industries (such as information service industry) support manufacturers, the cross-border integration and information interconnection of enterprises in different industrial fields are encouraged, high-end intelligent equipment is made domestically, the industrial Internet infrastructure construction is improved, and the development of industrial platforms is promoted systematically.

2.2.2 Strengthen Innovation Efficiency

Innovation is the engine of manufacturing development, and also the inexhaustible driving force for structural adjustment and optimization and the transformation of economic development mode. Therefore, both the United States and Germany have placed innovation in a paramount position in their strategic action plans for manufacturing development. But history has also shown that innovative attempts in technology, products, and processes often fail because they are either too fragile or divorced from the surrounding business ecosystem.

For example, Nokia's pioneering 3G mobile phone failed because its partners in its ecosystem were unable to develop video streaming, location-oriented devices, and automated payment systems in a timely manner. In 1980, the revolutionary high-resolution TV of Philips Electronics also ended in failure due to the lack of high-resolution cameras and supportive transmission standards.

An industrial platform can support innovation by strengthening the partnership between manufacturers and users. When platform users are developing new product ideas, other users can support this innovation by helping to develop a reliable supply chain, adopting relevant technologies, rendering related product services, and cooperating in the distribution and marketing of this new product.

2.2.3   Multiple Geographic Locations Create Economies of Scope

Companies that rely on traditional manufacturing methods can hardly serve any market that requires low-volume products. If a particular country or region cannot support production for a huge market, most companies will choose to ship products from another location (which may result in expensive products), or skip this market entirely.

The industrial platform has many solutions to this dilemma. The ability to electronically monitor or control small factories from afar, coupled with the flexibility to change production plans quickly and easily, enables manufacturers to build multi-mode factories even in low-density areas. By closely tracking demand, manufacturers can make precise decisions, change products or components according to demand, and make a variety of available products that are affordable even in small markets.

The industrial platform also makes it easier for multinational companies to track changes in the demand for a certain product across countries or regions. The newest information on consumer preferences and tendencies helps manufacturers modify production plans to meet demand more accurately, therefore reducing the cost of production, transportation and storage of unsold goods. This platform can also identify local resources of ideal components and raw materials, optimize the company's supply chain, lower the risks, and increase profits.

In addition, companies that share the same platform can combine orders for different products from the same region or country to create high-quality, high-value products within their own capability and affordability. The industrial platform can both track and combine orders, and identify the best shipping route and method; apply competitive prices within a certain period of time to generate more profits. These enhanced benefits generated by the industrial platform can help companies serve multiple and smaller markets and obtain higher profits than now.

Certainly, the construction of an industrial platform does not happen overnight. At present, from a practical point of view, the industrial platform is still at an early stage. Although platform technology and service capabilities have achieved single-point innovation, to make a system breakthrough, it is necessary to explore and build an open and cooperative ecosystem for win-win development.

2.3   Victory: the Pan-industrial Revolution

All manufacturing systems have two yardsticks, namely the manufacturing technology system and the manufacturing mode system that accompanies it. The manufacturing technology system is a collection of manufacturing technologies, and promotes the continuous development of productivity, while the manufacturing mode system affects the types of technology inputs, the nature of the conversion, and the output of the system. Therefore, the manufacturing mode system determines the effectiveness and efficiency of technology utilization. A simple attempt to optimize the technical system may reduce the total efficiency of social manufacturing.

The pan-industrial age is an age when the evolution of manufacturing technology synergizes that of manufacturing modes. At present, information technology represented by the Internet of Things and big data, new energy technology represented by green energy, and digital intelligent manufacturing represented by 3D printing technology are collaboratively innovating to integrate flexibility, intelligence, agility, lean-ness, globalization, and humanity. This will change the production mode of the manufacturing industry and the global economic system, leading people to live in the "pan" industrial age.

Only by grasping the "pan" industry, creating "pan" advantages, and overcoming the "pan" challenge can we have a victorious "pan" future.

2.3.1   Evolution of Manufacturing Technology

In the 1980s, the introduction of advanced manufacturing technology was the beginning of a new round of manufacturing technology evolution.

The concept of advanced manufacturing technology originated in the United States. In the early days, it referred to the manufacturing technology group based on computer and information technology. It mainly included computer-aided design, computer-aided manufacturing, computer-aided engineering, robotics and flexible manufacturing technology, automatic control system, numerical control technology and equipment, etc. However, as technologies advance, their connotation has been continuously expanded, including various new and advanced machining technologies, such as nano-processing, laser cutting, and additive manufacturing.

Judging from the flow of production, advanced manufacturing technology and traditional manufacturing technology are different stories of the manufacturing process. Traditional manufacturing uses manufacturing resources to convert raw materials into products. It is only a part of the production flow and generally includes two major steps: product processing and assembly. Manufacturers produce or purchase parts from suppliers, assemble them into products and inspect them to make sure they meet the requirements. The inputs in the manufacturing process are raw materials, energy, information, human resources, etc., while the output is products that meet the requirements. Information transmission and feedback between the various departments of system design, manufacturing, and sales in traditional manufacturing are not good enough. Each department breaks down tasks according to their functions. It is easy for them to only consider their own interests, and pay little attention to the optimization of the system, which leads to conflicts and difficulty to cooperate between the design and manufacturing departments.

Advanced manufacturing technology expands the traditional manufacturing technology from four aspects: material design, manufacturing process transformation, integrated solutions for product and service integration, and recycling. The molding and processing technologies of new materials are gaining more weight. And the directional transformation of the molecular or atomic layer of the materials has greatly enhanced product performance. Some new molding and processing technologies for super hard materials and functionally graded composite materials will continue to emerge, such as the molding and processing of superconducting materials.

For the transformation of the manufacturing process, traditional manufacturing is distributed processing that is oriented to batch processing and separated in time and space while advanced manufacturing super-efficiency processing and automation technology can keep the flow manufacturing going and reduce the inventory of components.

Besides, advanced manufacturing emphasizes the coverage of the entire process from product development to customer application. It provides product solutions and end-to-end services that integrate products, software, and services. The high degree of aggregation of intellectual capital, human capital and technological capital has rid manufacturing activities of the traditional manufacturing mode of low technology content and low added value. Through product design, management consulting and other activities, technology and knowledge are used in the production flow, thus technological progress is transformed into production capacity and competitiveness, and higher added value is generated for the enterprises.

In addition, advanced manufacturing also pays attention to the recycling of materials. It is not only environmentally friendly but also saves raw material costs. The traditional product manufacturing mode is an open-loop system, that is, raw materials → industrial production → product use → scrap → discarded into the environment. This is the manufacturing method that consumes a lot of resources and destroys the environment. However, advanced manufacturing technology considers the ecological environment and resource efficiency throughout the manufacturing lifecycle, and expands from the pure product function design to the life cycle design.

Among that, additive/precise manufacturing is used to transform the processing stage; robotics/automation technology is used to automate assembly and the production flow; advanced electronic technology is used to integrate products and services and control the processing; supply chain design aims at the best overall efficiency, and systematically considers the optimal combination of humans, technology, management, equipment, materials, information, and other system elements to meet the requirements of product or service supply at the lowest cost; cleaner production technology is mainly used for the recycling of materials; molecular biology and biological manufacturing are used for material design and manufacturing process improvement; nanomaterials technology is used to synthesize and process functionally graded materials, composite materials, etc.; the Internet of Things, cloud computing, and big data are used to track, analyze, optimize, and control the manufacturing processes of the whole life cycle of the product, realize multi-dimensional and transparent ubiquitous perception, and ensure the efficient, agile, sustainable, and intelligent manufacturing process.

Apparently, advanced manufacturing technology pays attention to economic benefits and technological integration. Through flexible production, product differentiation, focus on efficiency and quality, it enhances the enterprise's ability to respond to the market, improve independent innovation capabilities, and provide customers with better service. And it is characterized by excellent product quality, high technical content, low resource consumption, less environmental pollution, and satisfying economic benefits.

2.3.2 The Evolution of the Manufacturing Mode

The manufacturing mode is an effective production method / certain production organization mode that the manufacturing industry adapts to improve product quality, market competitiveness, production scale, and production speed so that specific production tasks are completed. From the perspective of the development of the manufacturing industry, different periods of social development determine different manufacturing ideas, production organization modes, and management concepts, which interact and jointly determine the manufacturing mode in a particular period.

High and new technologies represented by computer information technology and intelligent technology, with their high penetration, strong drive, and multiplier, not only led to the rapid rise of the information industry's and intelligent industry's positions in the national economy, but also reshaped the traditional manufacturing industry. They organize manufacturing activities in a flat, networked structure, pursue overall social benefits, a satisfying customer experience and corporate profitability, and become the most optimized flexible and intelligent manufacturing system. And the manufacturing mode with networked collaboration, personalized customization, service extension, and intelligent production has become the fruit of the synergy and deep integration of the new technology clusters.

From the perspective of networked collaboration, the manufacturing mode adopts information technology and network technology as the basis for rapid responses to market demand and enhances the competitiveness of enterprises in the manufacturing activities of a product's full life cycle. For example, enterprises use the Internet, big data, and industrial cloud platforms to develop new modes of collaborative R&D, crowdsourcing design, and supply chain collaboration among enterprises so that resource acquisition costs are effectively reduced, the scope of resource utilization is greatly extended, closed boundaries are broken, the transformation from flying solo to industrial coordination is accelerated, and the overall competitiveness of the industry is enhanced. Networked collaboration, which includes collaborative R&D, crowdsourcing design, and supply chain collaboration, opens up new channels for traditional enterprises to innovate at high-efficiency, great convenience, and low costs.

From the perspective of personalized customization, the Internet, big data, and cloud computing, algorithms, and flexible production capabilities and levels have improved, which has promoted the rapid development of personalized customization. With the help of industrial platforms, enterprises can have deep interactions with users, collect requirements extensively, perform big data analysis to establish production scheduling models, and rely on flexible production lines to provide personalized products while maintaining economies of scale. Personalized customization directly transforms user needs into production orders through industrial platforms and smart factories. It realizes user-centered personalized customization and on-demand production, effectively meets the diversified needs of the market, and solves the long-standing inventory and production capacity problems of the manufacturing industry, and ultimately reaches a dynamic balance of production and sales. For example, in the personalized manufacturing system represented by 3D printing, consumers no longer passively accept or choose their favorite products from the product list that enterprises give them, but participate in product design and directly make the product.

Service extension adds value to the manufacturing value chain. By integrating products and services, allowing customers to participate fully, and providing production-oriented services or service-oriented production, it integrates dispersed manufacturing resources and achieves the efficient synergy of their respective core competencies. Consequently, an efficient and innovative kind of manufacturing mode is created.

Enterprises can add intelligent modules to their products to realize product networking and operational data collection, and perform big data analysis to render diversified intelligent services, so that they expand from selling products to selling services, effectively extending the value chain and expanding profit margins. Building an intelligent service platform and intelligent service will become the core of the new business, so that enterprises can break free from the investment in resources, energy and other elements, and better meet user needs, increase added value, and improve overall competitiveness.

Intelligent manufacturing uses advanced manufacturing tools and network information technology to intelligently transform the production process, realize the cross-system flow, collection, analysis, and optimization of data, and implement the intelligent production methods such as equipment performance perception, process optimization, and intelligent production scheduling. Meanwhile, based on the new generation of information technology, cloud computing, big data, the Internet of Things technology, nanotechnology, sensor technology, and artificial intelligence, it intelligentizes product design, manufacturing, logistics, management, maintenance and service through perception, human-computer interaction, decision-making, execution, and feedback. It is the integrated collaboration and deep integration of information technology and manufacturing technology.

During product processing, intelligent manufacturing integrates sensors and intelligent diagnosis and decision-making software into the equipment. It upgrades equipment from program control to intelligent control, so it can adaptively feedback the condition of the processed workpiece in the process. For example, compared with traditional CNC machining technology, the intelligent manufacturing process based on CPS can perceive changes in temperature, environment, and properties of processed materials, and adjust accordingly. It will not rigidly execute predetermined programs but ensure the precision of processed products.

2.3.3 Win with the Pan

The emergence of new manufacturing technologies and manufacturing modes in large numbers with the transformation of production methods as the main thread, and the collaborative integration together push human society into the pan-industrial age.

Primarily, the Pan-industrial Revolution is a revolution that does not rely on a single discipline or a few types of technology, but on all-round multi-discipline, multi-technical level, wide-field synergy and deep integration. The new generation of information technology, including cloud computing, big data, the Internet of Things, the Internet of Services, cloud platforms, etc., new energy, such as renewable energy, clean energy, etc., new materials, such as composite materials, nanomaterials, and other technologies will constitute the strong and new infrastructure for the Pan-industrial Revolution. Decentralized manufacturing such as networked manufacturing, manufacturing IoT, cloud manufacturing, and smart manufacturing, crowdsourcing production, cluster effects, and niche thinking have revolutionized production methods, lifting the entire industrial production system to a new level. Industrial production, the economic system and the social structure will shift from vertical to flat, from centralized to decentralized.

In addition, the new generation of advanced manufacturing modes represented by intelligent manufacturing will certainly change business models, management models, service models, corporate organizational structures, and human resource requirements. This will fundamentally transform the industrial field, production value chain, business models and even lifestyles, thus promoting and realizing the Pan-industrial Revolution.

Secondly, the Pan-industrial Revolution will widely extend to all industries, whether it is the consumer-oriented Internet or aerospace, daily life or life science. It will be a multi-dimensional revolution embedded in the entire social system of technological economy. Therefore, the strategic preparation for it cannot be narrowly limited to breakthroughs in cutting-edge manufacturing technologies. It should integrate technologies and the industry to create a pan-industrial platform with innovation and management merits.

One is that manufacturing technologies are embedded in a larger technological innovation system. The Pan-industrial Revolution is the application of basic technologies including digital technologies, electronic technologies, and material technologies, as well as technical tools such as simulation, digital modeling, robotics, artificial intelligence, process control sensors, and measurement, throughout design, development, manufacturing, distribution, and service. Therefore, to achieve break-throughs in advanced manufacturing technology, countries must break the traditional, static technological and industrial boundaries, or cultivate independent innovation capabilities, or become able to access and utilize global innovation resources. Only by cultivating and integrating technical capabilities in various fields can competitive modern manufacturing capabilities be created.

The second is that manufacturing technologies are embedded in the enterprise's management system. The application and execution of the new generation of manu-facturing technologies have always been a systematic and coordinated transformation of manufacturing technologies and corporate strategies, marketing, and basic management. A survey on the American flexible manufacturing system revealed that up to 20% of the equipment in the flexible manufacturing system in the early 1990s sat idle, and what impeded their use was the lack of corporate management and employee capabilities that match with the new equipment. It is foreseeable that the Pan-industrial Revolution will inevitably come with changes in product innovation, management, and business models. The manufacturing revolution in the factories is only part of the overall strategic change of the enterprises.

The understanding of the connotation of the Pan-industrial Revolution must be through interdisciplinary dialogues and exchanges with social sciences (such as economics and management), so there is an appropriate breaking through the theoretical scope of natural sciences and engineering technology disciplines. The course of industrial development shows that the emergence of new production models is the product of interaction with specific social systems, organizational structures, and economic factors, and that new manufacturing modes will make new requirements on existing social systems and management methods, therefore furthering the transformation of enterprise management model and social system environment.

In general, driven by changes in the market, technology, socio-economic environments, and global integration, the manufacturing industry is experiencing a revolution of advanced manufacturing modes, one that implements advanced manufacturing technologies and revolves around fundamental changes in business methods. It involves comprehensive transformation in manufacturing concepts, manufacturing strategy, manufacturing technology, manufacturing organization, and manufacturing management.

PART 2
THE PRESENT

CHAPTER 3

Manufacturing Technology Clusters

3.1　3D printing: Additive Manufacturing and Precision Forming

In 2012, *The Economist* published an article in which 3D printing technology was commended as one of the major symbols of the Third Industrial Revolution, and it attracted great attention worldwide. As a revolutionary technology, 3D printing requires no operation in a factory, meaning that there is no need for machining or any molds. Without a doubt, this will greatly shorten the product development cycle, improve production efficiency, and reduce the costs of human resources required for production.

3D printing makes it possible for the public to fully participate in the entire manufacturing process of products, and for production and consumption models to be personalized, in real-time, and economical. As the advancement of 3D printing technology is accelerated, its application has gradually penetrated into many aspects of people's social life, and it is reshaping the social form and people's way of thinking and cognition in depth and breadth.

3.1.1 From Subtractive Manufacturing to Additive Manufacturing

Interestingly, 3D printing, which has only been widely known in recent years, was created as early as 1983. As a modern manufacturing technology developed on the integration of modern CAD/CAM technology, mechanical engineering, layered manufacturing technology, laser technology, computer numerical control technology, precision servo drive technology and new material technology, it is one of the important technologies that promote the new round of transformations in production models in the manufacturing industry.

3D printing, namely three-dimensional printing, differs from two-dimensional printing, but there are similarities, too. In fact, two-dimensional printing or three-dimensional printing is essentially a printing technology. The difference is that two-dimensional printing prints the content of the file in a plane shape. In addition to transmitting information, it has no actual functions. Compared with two-dimensional-printed files, 3D printing can directly achieve the functions.

3D printing first inputs the three-dimensional shape information of the object to be printed into a file readable to the 3D printer, and when it interprets the file, the three-dimensional shape is printed by stacking materials layer by layer. The three-dimensional shape is the basis of the function. When the shape is printed, the function is printed.

Also, in contrast to "subtractive manufacturing," 3D printing is also called "additive manufacturing." For the current manufacturing industry, most of the material processing technologies currently in use are "subtractive manufacturing" technologies, which remove, cut, and assemble raw materials, so that they have specific shapes and can perform specific functions. But "additive manufacturing" directly stacks raw materials layer by layer into specific shapes to achieve specific functions.

The course of work of additive manufacturing mainly includes two processes: three-dimensional design and layer-by-layer printing: first computer modeling software performs modeling, and then the built three-dimensional model is partitioned into layer-by-layer sections to guide the printer to print it layer by layer. Compared with traditional subtractive manufacturing, additive manufacturing certainly has many advantages.

First, it shortens the manufacturing time and improves efficiency. It usually takes several days to manufacture a model in the traditional way, depending on its size and complexity, while 3D printing shortens the time to merely several hours. Therefore, compared with subtractive manufacturing, additive manufacturing is particularly suitable for manufacturing parts with complex shapes. Certainly, this is also under the limitation of the performance of the printers and the size and complexity of the models.

Next, it enhances the utilization efficiency of raw materials. Compared with traditional metal manufacturing technology, additive manufacturing machines produce fewer by-products when manufacturing metal. With the advancement of printing materials, "net shape forming" manufacturing may become the more environmentally friendly processing method.

Thirdly, it completes complex structures to improve product performance. Traditional subtractive manufacturing methods have limitations in the processing of complex shapes and internal abdominal structures, while additive manufacturing can improve product performance by manufacturing complex structures. Therefore, in aerospace and mold processing, it has unparalleled advantages over subtractive manufacturing.

For example, a 3D printer can print many shapes, and it can make different components every time, like a craftsman. For traditional machine tool production lines, to process components of different shapes, complex adjustments to the production lines are required. Therefore, additive manufacturing is particularly suitable for customized, non-mass-produced items.

3.1.2 Smart Manufacturing Throughout Manufacturing

The application range of 3D printing is beyond people's imagination. In theory, it can print almost anything that exists. At the beginning, 3D printing was mostly used to make models in the fields of mold manufacturing and industrial design. As the technology matures, it has now been widely used in many fields, including aerospace, engineering construction, medical care, education, geographic information systems, automobiles, etc.

Traditional manufacturing: 3D printing is far better than traditional manufacturing technologies in terms of cost, speed, and accuracy. This technology itself is suitable for mass production. When the automotive industry conducts safety testing, some non-critical components will be 3D printed products, so as to reduce costs while increasing efficiency.

In August 2011, engineers from the University of Southampton in the UK designed and tested the world's first "printed" aircraft. Its outer shell was printed layer by layer by a 3D laser printer that specializes in nylon materials, and it turned out to be quite light.

In 2015, NASA 3D printed the head of an aircraft rocket engine. This greatly reduced the number of both components and weld seams. While lowering the probability of failure of the rocket engine, it also shortened the iteration cycle and reduced the cost.

In May 2020, a Long March 5B carrier rocket carried China's new manned test spacecraft with a "3D printer" on board. It was China's first 3D printing experiment in space and the first international 3D printing experiment of continuous fiber-reinforced composite materials in space.

Medical industry: In surgery, 3D printing can "tailor-build" the organs for patients who need organ transplants, without worrying about rejection. In addition, to print a human heart valve, it takes only US$10 worth of polymer materials. In 2019, Tel Aviv University announced that the school laboratory has successfully 3D printed a "heart."

It is not only a printed heart, but the first three-dimensional artificial heart with vascular tissues in the world printed by using the patient's own cells and biological materials.

Architectural design: In the construction industry, engineers and designers have begun to use 3D printers to print architectural models. It is fast, low-cost, environment-friendly, produces exquisite models that fully meet the expectation of the designers, and saves lots of materials. In Dubai, the government has even chosen to use 3D printing to build government offices. Machines finish the main tasks of 3D printing construction. One print does it all. So, the construction speed is fast. Workers mostly operate and inspect the working conditions of the 3D printers. Therefore, this technology requires less manpower than the traditional construction industry.

Personalized product customization: 3D printing technology successfully connects the virtual world with the real world. It transforms the ideas in people's heads into data modeling files in the computer, and turns them into real and sensible things through printing equipment. In the future, via the Internet, all products and consumables that people need in life can be printed. With matching materials and cloud service technology, each 3D printer can be real-time controlled, and produce the required items as required, whether it's a personalized pen holder, or a phone case with a half-body relief, or a unique ring of love.

At the same time, under the production method of social manufacturing, a large number of 3D printers will form a large manufacturing network. They seamlessly connect with the Internet and the Internet of Things and create a complex social manufacturing network system, whose prominent feature is that consumers can directly transform diverse needs into products, realizing the full conversion of "ideas to products." Consumers are able to participate in product design and improvement through the Internet, so that the diversified needs of society are met to the greatest extent. It is foreseeable that with the penetration of printing technology into different areas of social life, the human world will welcome a new age with a greater variety of products.

As technologies mature, 3D printing exhibits greater commercial value.

3.1.3   The Path Ahead for 3D Printing

As 3D printing enters people's production and lives, that everything can be printed is coming true. The industry looks forward to its huge development prospects and broad application space make the industry, but as a fast-developing technology, it still has a long way to go to harness the positive influence thereof.

On the one hand, for the processing of standard products, the scale merit of 3D printing is not as good as traditional processing methods. The fixed costs in the 3D printing manufacturing are less than those of traditional processing. This will lead to a reduction in the marginal cost of 3D printing manufacturing when standard products are produced on a large scale, thus less ideal than traditional processing.

For example, when traditional injection molding processes a rubber component, the mold used is a fixed cost. Since the product is standardized, when the component is processed in batches, the fixed cost of each component is reduced. Therefore, the number of components processed by the mold can be infinite, meaning the average cost of each part approaches 0; if the components are processed by 3D printing, no mold is required, so when it processes the same components in batches, there is no fixed cost reduction.

On the other hand, for 3D printing, there are limited types of raw materials currently available. Judging from the present situation, the types of materials qualified for 3D printing technology are not as large as those for traditional processing methods. There are two main reasons. First, for the raw materials of different properties, the applied equipment principles vary, so the development of raw material types for 3D printing is restricted to the progress of the corresponding equipment R&D; The other reason is that the raw materials for 3D printing often require specific forms. For example, metal 3D printing usually uses metal powders as the raw material, and has special requirements for their uniformity, oxygen content, and particle size. Compared with proximate matters, processing powder is more difficult, and the corresponding industrial chain is not as extensive and large as that of traditional materials.

For 3D printing that uses ABS plastics, photosensitive resins, and other non-metallic materials, there are many raw material suppliers on the market, and their cost is no longer a bottleneck restricting the development of this technology; but for metal and high-end polymer materials, due to the limitation of supply capacity, the price is still relatively high.

In addition, there is still room for improvement in the mechanical properties of 3D printed components and metal 3D printing in terms of processing accuracy, surface roughness, and processing efficiency. Meanwhile, whether the finished product is robust and durable, whether the user's awareness is improved, and whether the intellectual property rights are exposed to more risks of infringement are all the must-walk paths ahead for 3D printing during its development.

3.1.4   The Imagination of Future Manufacturing

From the perspective of the commercial application and marketization of 3D printing, after over 30 years of development, the 3D printing industry has built a relatively complete industrial chain, including upstream components required for manufacturing 3D printing equipment, various raw materials used in printing, software and hardware necessary for design, and reverse engineering; midstream 3D printing equipment and services; downstream applications in aerospace, automotive, medical care, and education.

In fact, the scale of the 3D printing industry has shown rapid growth in both the global market and in the Chinese market. According to statistics from the consulting firm Wohlers Associates, the total output value of the global 3D printing industry in 2013 was US$3.03 billion, and in 2018 it reached US$9.68 billion, with a five-year growth rate of 26.1%. The firm also predicted that from 2019 to 2024, the global 3D printing industry will maintain an average annual growth rate of about 24%.

Compared with the global average, the Chinese market of 3D printing is growing faster. In 2013, the scale of the Chinese 3D printing industry was only US$320 million, and in 2018 it rose to US$2.36 billion, with a 5-year growth rate of 49.1%. It is estimated that in 2023, the total revenue of the Chinese 3D printing industry will exceed US$10 billion.

Clearly, 3D printing has gradually moved from the introductory phase to the growth phase, and COVID-19 has undoubtedly accelerated this process. 3D printing is not subject to space, can shorten the supply chain process, and increase production efficiency. Moreover, it has a lower manufacturing threshold, too. Therefore, it gives a feasible solution for getting rid of the supply chain. As a result, under the shortage of overseas medical protective goods and materials, masks, medical masks, and goggles made via 3D printing save the day.

The Chinese consumer-level 3D printer manufacturer Creality is both a beneficiary of this wave of demand for 3D printing, and a major participating force in the domestic 3D printing industry. Creality shipped over 50,000 units in March, received orders for nearly 160,000 units in early April, and achieved sales of RMB220 million in April. In addition, export-oriented 3D companies such as Flash Forge, iYong Technology, ESUN China, and FLSUN 3D also reported positive news.

In fact, many technological innovations are on the edge of breakthroughs, but few are expected to reverse the decline in productivity growth. However, 3D printing is an exception—it is designed to be a tool to increase productivity. If 3D printing and robots are combined, it can cast a greater impact. Robots are flexible in 3D space while 3D printers can build complex objects. Combining the two means the possibility to build any structure from scratch.

In the past four decades, China's manufacturing industry has experienced rapid development from recovery to prosperity. It continues to expand its total scale, accelerate the transformation of its industrial structure, and significantly increase its overall strength and international competitiveness. Today, as the manufacturing industry evolves quickly to Industry 4.0 smart manufacturing, the biggest challenge of smart manufacturing has transitioned from quantity to quality. Changes like mass customization, open innovation and smart factories will also be the most direct manifestation of 3D printing smart manufacturing.

3D printing technology carries people's imagination of the future manufacturing mode. It is a new technology bred from the accumulation of technologies in the digital age to a certain stage. It allows humans to imagine a promising future. In the future, the physical restrictions and space limitations of traditional manufacturing will no longer get in the way, and the design and production will be flatter and more open.

3.2   Nanotechnology: Encyclopedia on the Head of a Pin

In December 1959, physicist Richard Feynman gave a lecture entitled "There's Plenty of Room at the Bottom: An Invitation to Enter a New Field of Physics" on the theme "Manipulating and Controlling Things on a Small Scale." In that lecture, Feynman was not satisfied with the technique of engraving letters on the needle, and further proposed: "Why cannot we write the entire 24 volumes of the Encyclopedia Britannica on the head of a pin?"

It was this idea which seemed far-fetched at the time and failed to attract too much attention that became the earliest scientific prediction of nanotechnology and fundamentally opened the curtain to its conscious scientific development. In 1990, Don Eigler and Erhard Schweizer used a scanning electron microscope to manipulate a single xenon atom on the nickel surface. They managed to manipulate the atom to write "IBM" for the first time, realizing Feynman's vision.

At present, after decades of development, nanotechnology has become common. As an innovative technology on the micro-nano scale, it can produce highly flexible, conductive, and durable new materials. The nano-instruments used and the nanoparticles prepared have also brought about significant changes in various fields of science, industry, and daily life.

3.2.1   From Length Unit to Technical Possibility

As a unit of length, the nanometer is no stranger. One nanometer is one-billionth of a meter. A molecule and DNA are One nanometer, a hair 75,000 nanometers, the needle used for injection, one million nanometers, and a two-meter tall basketball player, two billion nanometers.

When a certain dimension of the three-dimensional size of the material reaches the nanometer level, between 0.1 and 100 nanometers, this material can be called a nanomaterial. Moreover, on the nanoscale, the material will exhibit totally different physical, chemical, and biological properties from the macroscale.

For example, in a conventional chemical reaction, bonds combine atoms, the reactants are kept in the precise direction that promotes the lowest free energy, and each reactant has discontinuous energy. The rearrangement of atoms in a chemical reaction always comes with the release or absorption of heat. Breaking bonds absorbs energy while forming bonds releases it.

When carrying out the same reaction in the nanoscale, the reactants can maintain a precise direction through the clamps on the conveyor belt using a "molecular machine," and they can be held together at an appropriate angle and strength. The conveyor belt moves as the reaction takes place, catalyzing over one million reactions per second.

In fact, the different characteristics exhibited on the nanoscale have been recorded in ancient times. However, despite the ancient applications, they did not form a discipline. In the 6th century B.C., the Lycurgus Cup was made by ancient craftsmen by adding colloidal gold and silver nanoparticles to the materials. Under the action of light, the color of the cup could change from green to red. The Lycurgus Cup, now collected in the British Museum, is the earliest nanotechnology application discovered so far.

Obviously, the shrinking of scale makes nano-materials present strange properties different from both macroscopic materials and individual isolated atoms. Based on this scientific discovery, nanotechnology was born.

Nanotechnology is a technology for exploring and controlling substances below 100 nanometers, manipulating atoms and molecules in a certain space, processing materials, and manufacturing devices with specific functions. The truth is that molecules designed for specific functions have always been part of modern chemistry. But nanotechnology is not chemistry. It is not subject to the attraction and combination of molecules and ions in the solution.

That is to say, once the "bottom-up" specific steps (creating an atomic-level precise structure) are worked out, the design of new nanomachines and nanofabrication systems will be similar to mechanical engineering—applicable to both single small components and to large systems.

The scientific and technical revolution has provided new tools for human production and life. Nanotechnology uses precise operations at the nanoscale to regulate the properties of substances and endow nanomaterials with ideal mechanical, chemical, electrical, magnetic, thermal, or optical properties. New nanomaterials are widely applied in traditional and new industrial manufacturing.

3.2.2 Nanotechnology in the Industry

At present, after decades of development, nanotechnology has become common. It has provided an innovative driving force for the seven basic disciplines of physics, materials science, chemistry, energy science, life science, pharmacology and toxicology, and engineering, becoming an important source of transformative industrial manufacturing technologies.

In the medical field, nanotechnology has provided medical staff with new non-invasive nanomedicines, making significant progress in the treatment of some of the most difficult diseases. It provides new ways for drug delivery and disease treatment. With the help of nano-carriers, drugs can overcome the biological barrier of the human body and directly reach the focal zone via artificial manipulation, thus increasing the local drug concentration and enhancing the therapeutic effects while reducing damage to other tissues. It has demonstrated its advantages in cancer treatment.

Diseases such as multiple sclerosis, Alzheimer's disease, and Parkinson's disease may be effectively controlled with simple injections twice a year. These new nanomedicines can popularize healthcare and make it more affordable. It can also transform healthcare through new clinical applications such as drug delivery methods, mobile diagnostics, new therapies, nano-vaccines, nano-stents, anti-microbial treatments, implants and prostheses. Technological advancement in sensor networks, nanomaterial optics, and the Internet of Nano Things (IoNT) can also enable doctors and even patients to better control the diagnosis and treatment process and monitor their health through remote medical treatment and remote surgeries.

Epidemics and plagues can be monitored and controlled using nanosensors, new instant diagnostic methods, and nanomedicine. There have been studies on the use of nanotechnology-related treatment and diagnostic tools to prevent viruses such as Ebola and Zika from infecting the entire world. Nanotechnology will enable us to better face these large health threats.

It is predicted that by 2050, the overuse of antibiotics by humans and animals will have become a more widespread health problem than cancer, and nanotechnology can extract tiny particles such as antibiotics or plastics from food and water. Nanomedicine is studying the use of highly targeted non-invasive treatment options to cure various cancers.

In construction, building composites with nano-components will cast a positive impact. Different types of nanomaterials can also be used in binding materials such as concrete and cement to improve their performance. For their high tensile strength, nanomaterials, as sensors, can be used as pressure gauges and strainmeters during and after construction to ensure a reasonable structure of the building, thereby solving the load-bearing and aging problems of the building under extreme temperatures and making urban development more sustainable. In addition, nanotechnology can make buildings resistant to harsh climate and environmental pollution (including the absorption of carbon dioxide in the air). While making the construction process easier, it allows the buildings to better resist the dangers of earthquakes and tsunamis that may strike in the future.

Nano-sensors and graphene batteries that can be built into the road have been invented. They can charge electric vehicles while managing traffic flow and transmitting traffic data to achieve the best use of time and energy. Although graphene-incorporated nano-level roads are not yet popular, a one-kilometer-long road test has begun in the suburbs of Rome, Italy. The test shows that adding a small amount of graphene to the asphalt can improve the wear resistance of the road, thus expending its lifespan by 6 to 12 years, and withstand climate changes. Moreover, when nanotechnology is applied, urban road signs can store data, thus building a safer traffic environment for future smart cities and their citizens. China has begun to apply nano-emulsion ink to road signs to track traffic conditions across the country, which will greatly improve the traffic environment, safety, and traffic flow.

The application of nanomaterials in industrial manufacturing has a wide and profound impact on the present and the future. For example, in new energy, nanotechnology has brought new opportunities for the development of lithium batteries. It solves the major issues such as the safety of charging and discharging of the traditional lithium batteries (using silicon nanowires or S/nano TiO_2 with a hollow shell), slow speed (using carbon nanotubes), and battery instability (using super two-dimensional BN/graphene composite).

In fact, the ongoing research on nanomaterials for lithium batteries has been perfected and industrialized. The energy density of commercial lithium batteries has exceeded 300 Wh/kg, enabling the endurance mileage of lithium battery-powered vehicles to be up to 470 kilometers. With the further development of nanomaterials and the further optimization of lithium battery performance, the energy density is expected to reach 500 Wh/kg, thus achieving an endurance mileage of 800 kilometers.

A friction nanogenerator is a new application of nanotechnology in mechanical energy power generation. It can collect mechanical energy that is not easily obtained by traditional generators, such as friction energy, wind energy, ocean wave energy, and mechanical vibration. At present, some products using nano friction generators have been launched, such as self-powered smart shoes and triboelectric air purifiers. As a new technology that supplies green energy, friction nanogenerators will provide a new solution to the micro-energy supply problem in the development of the Internet of Things, as well as a new technical solution for large-scale "blue energy" (ocean energy). There are broad application prospects in the power supply of universal portable electronic products. Friction nanogenerators, which apply nanotechnology, may also lead technological innovation and profoundly change human society.

In the electronic information industry, the application of nanotechnology will help tear down the physical barriers represented by the strong field effect and quantum tunneling effect, and the technical barriers represented by power consumption, heat dissipation, and transmission delay, create a quantum effect-based new nano-device, and promote the development of cost-effective manufacturing processes.

In light industry, common sunscreen is mainly composed of nano-titanium dioxide or zinc oxide, and nano-fibers are used to make anti-wrinkle, anti-staining, and antibacterial clothing, as well as various sporting goods such as tennis rackets and bicycles.

Although nanotechnology is emerging as a cutting-edge technology, it is not far away from people's lives. It can be said that it has become part of people's lives without being noticed. For example, sunscreens usually contain nanoparticles of titanium dioxide and zinc oxide, both of which are excellent UV absorbers.

Obviously, nanotechnology can innovate on the micro-nano scale to create new materials with excellent flexibility, conductivity, and durability. The nano-instruments used and the nano-particles prepared have changed science, industry, and daily life. In the ever-developing age of science and technology, the impact of nanotechnology on people's production and life is far from the end.

3.2.3 Nano Layout Inspires Nano Thinking

Based on the broad application of nanotechnology in the future, all countries are constantly making nanotechnology strategies and actions.

In 2000, the United States took the lead in issuing the National Nano Plan to strengthen its coordination in the development of science, engineering, and technology at the nanoscale. In the past 20 years, in addition to maintaining high investment in nanotechnology basic research and infrastructure, the United States has paid more heed to the development of nano-enabled projects (NEPs), that is, the use of nanotechnology to develop materials, devices, and systems to support the upgrading of traditional industries and emerging industry applications.

Also, almost all industrialized countries in the world have accelerated the pace of advancing nanotechnology strategies and research plans. Newly industrialized countries and developing countries such as South Korea, Russia, China, Vietnam, and Israel have also formulated a series of nanotechnology development strategies and plans based on their respective national conditions.

In China, in 2001, the Ministry of Science and Technology and several national departments co-issued the National Nano Science and Technology Development Program and established the National Nano Science and Technology Steering and Coordination Committee. It was proposed to strengthen basic research, overcome key technologies, and cultivate key talents. Various national ministries and commissions have supported the R&D of new nano materials and new technologies through the National 973 Plan and National 863 Program.

In 2013, the Chinese Academy of Sciences launched the "Nano Pilot Project," hoping to use nanotechnology to promote transformative innovations of industrial technologies, such as long-lasting lithium batteries and nano-green printing. It is also expected to widen the application of a batch of core nanotechnologies in specific fields, such as energy, environment, and health, so as to solve a number of key technical bottlenecks restricting the development of Chinese backbone industries and encourage the development of emerging industries.

In 2016, the Ministry of Science and Technology issued the "Thirteenth Five-Year National Science and Technology Innovation Plan." It was focused on the R&D of major special projects of new nano-functional materials, nano-photoelectric devices and integrated systems, nano-biomedical materials, nano-medicine, nano-energy materials and devices, and nano-environmental materials.

Under the support of various projects and plans, China's nanotechnology is going great. China has become a strong power in nanotechnology R&D. There have been a series of achievements, including the successful development of a process platform for integrated circuit technology R&D of 22 nanometers and below; the establishment of the world's first production line to achieve large-scale and low-cost production of high-quality graphene; the launch of the first fully automated production line for the mass production of graphene organic solar optoelectronic devices; and the trial production of 40 nm and 28 nm system-level chips.

In fact, in addition to the strategies and actions of nanotechnology affecting the production and life of society at the technical level, more importantly, nanotechnology can be derived to the methodological level—nanotechnology inspires nano thinking, which, from the perspective of nanosizes, nanofeatures, nanotechnology, considers repositioning the boundaries of products, and inspires new ideas and solutions for product upgrades and industrial upgrades.

As Internet technology inspires Internet thinking and changes many models of traditional industries, nanotechnology will further change the quality and performance of products. Every nanometer shortened means the re-selection of material technology, the adjustment of supporting systems, and a long-term reference standard.

At present, after decades of development, nanotechnology has become common. From nanotechnology to nanothinking, human society will make new progress.

3.3   New Energy Technology: Green and Sustainable

"To change the burning of carbon-based fossil fuels to using renewable new energy, to re-understand what elements make up the world, to transform every building into a mini energy harvester that collects renewable energy on-site; to store hydrogen and other storable energy in the buildings, to utilize all the social infrastructure to store intermittent renewable energy, to ensure a durable and reliable supply of eco-friendly energy; to apply network communication technology to transform the power grid into a smart universal network, so that millions of people can transmit the electricity generated by surrounding buildings to the power grid, and share resources in an open environment. The operating principle is the same as the generation and dissemination of information on the Internet; to change the global transportation mode consisting of automobiles and trains into a transportation network composed of plug-in and battery-burning vehicles powered by renewable energy, and to build charging stations nationwide, so that people can buy and sell electric energy there."

As the famous American futurist Jeremy Rifkin put it, renewable energy technology will inevitably lead us to a green and low-carbon industrial age. The development of new energy technology will become the key to building a green future.

3.3.1   Time for Energy Conversion

Although modern people mostly use coal and oil to define the technological revolutions in the first and second energy fields of human civilization, they believe that the two have had a revolutionary impact on energy production and energy consumption. But on a larger time scale, vegetation energy was the start for the development of human civilization. Obviously, human beings used to rely on that for most of their lives and thus have reached the pinnacle of farming civilization.

At the beginning of human civilization, its development relied on the materials that can be directly obtained in nature and used for consumption, such as plants and animals. Primitive humans used to dwell in either natural shelters or natural places that could be modified into shelters, like caves

Vegetation has continuous reproduction that breeds in an endless succession, but the reproduction is subject to the region and its number has an upper limit. In this age, people live where there is water and vegetation. A village or a city needs vegetation with 30–50 times the size of the settlements to support the daily energy consumption there.

In the stage of vegetation energy, humans converse energy in an extremely simple fashion. The understanding of the relationship between "fire" and themselves is reflected in an important energy manipulation technology that the ancients invented. From the fear of wildfires in forests or on grasslands caused by lightning strikes to learning to use fire to grill prey, and to using it to keep out the cold, illuminate, and keep away wild beasts, the mastery of artificial fire making marks fire as a natural force that was truly utilized by human beings.

More than two hundred years ago, James Watt invented the steam engine. They burned coal to turn water into steam, and replaced human labor to handle cumbersome and laborious work, thus starting the Industrial Revolution. The epoch when fossil fuel was the main energy source began. In those years, the so-called first energy revolution and the second took place.

Among them, the UK represented the first energy transition, where coal replaced the dominant firewood. According to Vaclav Smil's quantitative standards, the UK's energy transition began in 1550 and ended around 1619, taking around 70 years.

Around 1550, the proportion of coal in the UK energy consumption structure exceeded 5%; around 1619, the number surpassed that of the dominant firewood, completing the transition to the coal system. Once that was completed, as economic development and the Industrial Revolution marched forward, the proportion of coal in the UK energy consumption structure continued to grow, reaching a historical peak (97.7%) in 1938.

The U.S. led the second energy transition, where oil replaced the dominant coal. According to Vaclav Smil's quantitative standards, the energy transition in the U.S. began in 1910 and ended in 1950, taking only about 40 years. In 1950, the proportion of oil in the U.S. energy consumption structure (38.4%) surpassed that of coal (35.5%) for the first time, becoming the dominant energy source.

In the past few hundred years, the use of fossil fuel and vigorous technological innovation have allowed mankind to enjoy unprecedented prosperity and abundance. The world population and GDP per capita have skyrocketed. The productivity created by mankind in less than a century is greater than that created in all past years, because industrial machinery, chemistry, ships, railways, and telegraphs all require substantial energy as the basis and support.

Judging from the economic development speed and energy supply curve of the past few decades, there is a close tie between the economic development of modern society and energy. The fluctuation of energy supply inevitably leads to the fluctuation of economic development, and vice versa. Mankind cannot survive without energy, and the destructive consequences from energy supply interruption clearly demonstrate the dependence of mankind's basic production and life on energy.

However, behind the heavy dependence on fossil energy lurks a serious crisis: on the one hand, the exploitation of fossil energy has an end. Despite the existence of undiscovered fossil energy reserves, they are not unlimited after all. Should no suitable alternative energy be found, as per the consumption rate in 2018, the global fossil energy will be exhausted in about 80 years. On the other hand, large-scale exploitation and utilization of fossil energy have led to increasingly severe environmental and climate problems. At present, people mainly use fossil energy via direct combustion. The sulfur and nitrogen contained in it are discharged into the atmosphere and form corrosive pollutants such as acid rain. Meanwhile, other pollutants such as smoke and dust are emitted during its development, production, and utilization, thus damaging and polluting local natural environment and geology. A large amount of carbon emission during the use of fossil energy is the culprit of global warming. A large amount of carbon is originally stored in the fossil energy in the earth's rock layers. During its combustion, carbon is released into the atmosphere as carbon dioxide, which rapidly increases the carbon dioxide content in the atmosphere and causes the earth's temperature to rise, i.e. global warming. These problems will have a serious impact on the earth's ecological environment, and ultimately challenge the development and survival of mankind.

Therefore, sustainable new energy sources, such as solar energy, geothermal/ocean heat, and natural mechanical energy (wind energy, tidal energy, and others), have to gradually replace fossil energy and become the main force that supports the operation of the society and people's survival. The next energy revolution will also start from this.

3.3.2   Dilemma in the Energy Transition

Whether it is the EU's "Energy 2020" project announced in 2010 that chose the green energy path, or the Japanese government's 2015 "National Rejuvenation Strategy" that made it clear to re-emphasize nuclear energy; whether it's the U.S. government's "Comprehensive Energy Strategy" in 2014 that highlighted occupying the commanding heights of the world's energy technology in the future, or the Indian government's announcement of the large-scale development of green energy in 2015, energy issues have gained more importance and become the core strategic issues of major countries.

Energy has never been prioritized as much as today. When people are faced with an energy iteration again, a new energy age is coming at a faster pace. The key to the energy transition is the ability to develop and use new energy sources on a large scale. New energy sources refer to wind power, solar energy (including photovoltaic, solar thermal, and thermal power utilization mode), biomass energy and ocean energy.

To realize the energy transition, we must first ensure the technical and economic feasibility of the development and utilization of energy. Only when there is technical feasibility can energy be developed and utilized; only when there is economic feasibility can it be promoted and applied sustainably, and the unit energy cost should be affordable when environmental costs and other factors are taken into account. However, now, despite people's awareness of the challenges from the energy problem and actions to seek solutions for future energy supply, neither energy technology nor energy governance has yet to find a good solution.

From the perspective of energy technology, we still need to overcome multiple technical dilemmas: with a certain amount of resources, the only way to greater supply capacity of new energy is to improve the efficiency of energy conversion through advanced technology. For wind power, the key is to change the technical route of converting wind power into electrical energy through wind wheels, so as to break through the conversion efficiency limitation of Betz theory, and simultaneously reduce the difficulty of manufacturing wind power conversion equipment. For solar photovoltaic utilization, the challenge is how to continuously improve the efficiency of solar photovoltaic conversion. The conversion efficiency of advanced photovoltaic power generation in current commercial applications is about 25%. In theory, this efficiency can be increased to exceed 70%. Therefore, there is room for improvement.

In addition, new energy sources are characterized by being intermittent and uncertain, which contradict the requirements of continuous, reliable, sustainable and stable energy supply. Therefore, when developing new energy sources, it is necessary to develop supporting energy technologies, the most important of which are large-capacity energy storage technology, and energy sources that complement or retro-regulate new energy sources.

From the perspective of energy governance, the energy issue is not the issue of one single nation. All nations are in this together. The fate of all living things on earth ultimately depends on the choices of mankind. The problem is that although mankind has realized this problem, how to abandon differences and achieve cooperation for a unified global action remains a conundrum. There are still serious problems that impede effective energy governance.

Second, energy will not cast negative external effects to the environment, atmosphere, etc. during the development, transportation, and use of energy. The energy process will inevitably affect the external environment, but this effect is either positive, or within an acceptable range of negativity, or can be corrected through technical and management measures.

At last, this kind of energy can be safely developed and utilized on a large scale, so that it is able to replace traditional energy. Judging from the current state of energy technology exploitation, although marine energy use remains experimental and small-scale, the applications of hydroelectric power, nuclear energy, wind energy and solar energy have been relatively mature, and made achievements in the past two decades. However, the proportion of these types of energy in total energy consumption is still less than 16%. Moreover, no single energy source accounts for more than 10% in total energy consumption.

3.3.3 Build an Energy Technology System

Energy technology is a huge system that requires renewable energy to be the main body, terminal energy to be mainly electric energy, and multi-energy and multi-networks to integrate and complement each other. At present, energy technology can be vertically divided into nine fields of coal, oil and gas, nuclear energy, hydro energy, wind energy, solar energy, biomass energy, energy storage, integration of smart grids and energy. Horizontally, it includes three levels: innovative technology, forward-looking technology, and game-changing technology.

The coal sector should focus on high-efficiency coal combustion technology, coal power waste control technology; terminal scattered coal utilization technology, carbon dioxide capture, transmission and utilization technology; magnetic fluid combined cycle power generation technology.

The oil and gas sector should focus on full-wave seismic detection technology, precise steering and intelligent drilling technology; intelligent well completion and oil extraction technology; bionic drilling and extraction system technology.

The nuclear energy sector should focus on advanced deep uranium resource development technology; pressurized water reactor optimization and large-scale promotion and utilization technology; fast reactor and fourth-generation reactor development and utilization technology; nuclear fuel cycle front-end and back-end technology matching; modular small reactor multi-functional application; R&D of controllable nuclear fusion technology.

The hydro-energy sector should focus on high-head and large-flow hydro energy technology, dam construction technology; environmentally-friendly hydro energy utilization technology, large-scale maintenance technology; intelligent design, intelligent manufacturing, intelligent power generation and integrated technology of intelligent watersheds for hydropower stations.

The wind energy sector should focus on wind energy resource assessment and monitoring, high-power wind turbine design; wind turbine operations and maintenance, and fault diagnosis; high-altitude wind energy technology for high-power wireless transmission.

The solar energy sector should focus on crystalline silicon battery upgrades, solar thermal power generation; thin-film battery technology, solar hydrogen production technology; wearable flexible and lightweight solar battery technology.

The biomass energy sector should focus on the co-processing of urban and rural waste and poly-generation; the preparation of biomass functional materials; the selection, breeding, and planting of energy plants.

The energy storage sector should focus on lithium battery technology with high energy ratio and safety, lead-carbon battery technology with high cycle times; liquid sodium-sulfur battery technology; lithium-sulfur battery technology, and solid oxide electrolytic cell (SOEC) electrolytic hydrogen energy storage.

The integration of the smart grid and energy grid sectors should focus on technology to improve long-distance transmission capacity, high-proportion new energy absorption technology, and large-scale grid automation technology; high-efficiency energy conversion technology, transparent grid/energy grid technology; intelligent equipment based on functional materials, intelligent equipment based on biological structure topology, ubiquitous network and virtual reality (AR).

As technologies in various energy sectors are deeply integrated, the fuel conversion system can convert coal to gas, coal to oil, biomass to diesel, and biomass to natural gas, thus supplementing oil and gas resources. Coal, natural gas, and wind and sunlight make up a multi-source-combined heating and cooling system as well as a hydrogen production system. When there is abundant wind power and photovoltaic power, electric energy is converted into other forms of energy. Meanwhile, the decarbonization and limpidity via coal-to-hydrogen production will integrate wind energy, hydro energy, photovoltaic energy, thermal power generation, and energy storage to achieve a cascaded use of energy.

The internal logic of the energy revolution is the drive to push forward the development of human civilization – primitive society energy mainly met survival needs; the quality of human life in feudal society was improved, and primary industrial production greatly increased the demand for energy; the development of social civilization has accelerated since the Industrial Revolution. Human beings showed greater demand for transportation, information, culture, and recreation, and the modern industry's demand for energy has reached an unprecedented peak.

As waste water, waste gas and waste residues generated during the development and utilization of high-carbon energy give rise to a series of ecological and environmental problems, the ecological demand for energy production and consumption has once again entered the stage of new energy development. Like it or not, the energy replacement that is happening is the inevitable trend, and there is no alternative.

3.4 Industrial Robots: the Crown Jewel of Today's Manufacturing

Robots are the soul of contemporary industry.

Reviewing the development track of modern industrial manufacturing is enough to understand the significance of the machines to the processing and manufacturing industry. In 1784, the birth of the steam engines was a milestone in the First Industrial Revolution. The reliable utilization of steam led to a new generation of steam-powered engines and ignited the First Industrial Revolution. Combining industry, technology, and trade, "Science and Technology + Industrialization" paved the way for the success in the industrial age.

Similarly, the combination of science and technology and industrialization today will also promote the development of the new industrial age. Among the emerging technologies characterized by informatization and digitization, industrial robots should not be underestimated.

3.4.1 The Biggest Contribution of Industrial Robots: Greater Productivity

Industrial robots are industrial-oriented multi-joint manipulators or multi-degree-of-freedom mechanical devices that can perform tasks automatically. As a machine that relies on its own power and control to achieve various functions, it can be commanded by humans, run in accordance with pre-arranged procedures, or act in line with the principles and guidelines formulated by artificial intelligence.

In fact, since ancient times, mankind has been studying ways to reduce the workload, and to make work more convenient and efficient without damaging the quality. As early as the Western Zhou Dynasty over 3,000 years ago, puppets that could sing and dance appeared in China. They may be the earliest "robots" in the world.

Since modern times, along with the first and second Industrial Revolutions, the invention and application of various mechanical devices, industrial robots were ready to come out. In the 1950s and 1960s, as mechanism theory and servo theory were developed, robots were put to use. In 1960, AMF produced the cylindrical coordinate robot Versatran, which could be used for point and trajectory control, becoming the world's first robot for industrial production.

In the 1970s, with the advancement of computer technology, modern control technology, sensor technology, and artificial intelligence technology, industrial robots evolved rapidly. During this period, the robot had memory and storage capabilities, and could repeat operations according to corresponding procedures, but basically has no perception and feedback control capabilities for the surrounding environment. This kind of robot was known as the first generation robot.

In the 1980s, thanks to the progress of sensor technology, including visual sensors, non-visual sensors (force, touch, proximity, etc.), and information processing technology, the second generation of sensory robots were created. They were able to obtain some information about the work environment and work objects, perform a certain degree of real-time processing, and guide themselves to perform tasks.

The third-generation robot is the "intelligent robot" of today. They can both better perceive the environment than the second-generation robots, and perform logical thinking, judgment, and decision-making, thus able to work autonomously according to job requirements and environmental information. As an important part of intelligent manufacturing, they are a major weapon to realize intelligent production and build intelligent factories.

Industrial robots are a key link of intelligent manufacturing. For industrial robots, it is the first priority to be able to assist in solving the problems during the manufacturing. In other words, as machines have improved productivity in all past Industrial Revolutions, the greatest contribution of industrial robots lies in the improvement of manufacturing productivity rather than that of the robots themselves.

This is because the intelligent part of industrial robots can be regarded as an "agent" —tasks are assigned to the bottom level of the control system for processing, and sensors, visual images, logic control, and communication work together to build a streamlined and effective control system for the bottom level (or called the core layer). Numerous "agents" in the system communicate with each other and generate a swarm intelligence.

This swarm intelligence can be used in a variety of production activities, in different single product production lines, or in different production scales, including some flexible production lines. Applying industrial robots to industrial production lines not only improves production efficiency, but also the working environment. While ensuring the safety of workers, it can reduce the loss of raw materials and lower industrial costs from the source.

Since Germany first proposed the "Industry 4.0" strategy with intelligent manufacturing as the core in 2011, the Fourth Industrial Revolution marked by intelligence has swept the world. Along the process, as a perfect combination of industrialization and informatization, industrial robots, with their characteristics of digitalization, have connected a single production equipment to the entire production network, thus supporting the rich application scenarios of the Fourth Industrial Revolution.

If the development of the Internet in the past two decades is considered what has connected each of us, the development of the industrial Internet in the next two decades will connect every industrial robot, thereby comprehensively transforming production efficiency and production methods.

3.4.2 Intelligentize Manufacturing

Industrial robots can be divided into many types according to different classifications. According to the mechanical structure, there are tandem robots and parallel robots. The movement of one axis of a tandem robot changes the coordinate origin of the other axis, such as a six-joint robot. The studies on tandem robots have been more mature.

With the merits of simple structure, low cost, easy control, and large movement space, they have been successfully applied in many fields, such as various machine tools, assembly workshops, etc.

Parallel robots can be defined as a parallel-driven closed-loop mechanism with two or more degrees of freedom that is connected by a moving platform and a fixed platform through at least two independent kinematic chains. Generally, 3-axis parallel robots are the most common. Parallel robots are characterized by zero cumulative error and high accuracy. Their driving device can be placed on or close to the fixed platform, so that the moving part has light weight, high speed, and good dynamic response.

Parallel robots are generally used for sorting, handling, boxing, labeling, and testing of light and small objects on the production line. They are widely used in food, pharmaceutical, electronics, and daily use chemical industries. At the beginning of their advent, their application objects used to be mainly large dairy companies and pharmaceutical companies that produce liquid medicines in bags and tablets. Most of the application load was below 3 kg. Subsequent growth mainly comes from the food industry outside the dairy industry, such as candy, chocolate, moon cake manufacturers, and medicine, 3C electronics, printing, and other light industries.

According to the polar coordinate form of operation, they are divided into cylindrical coordinate robot, spherical coordinate robot, multi-joint robot, plane joint robot, etc. Among them, the joint robot is also known as the articulated robotic arm. As one of the most common forms of industrial robots in the current industrial fields, it is suitable for many industrial mechanical automation operations. According to the number of axes, it has multiple types. At present, four-axis and six-axis robots are most widely used. And the six-axis robot has six freely rotatable joints, which allow it to move freely in three-dimensional space and simulate the actions of human hands. With extremely high versatility, it is the most widely used, but most difficult to control and most expensive.

With different end effectors, the articulated robot can perform different functions. The high degree of freedom allows it to flexibly bypass the target and perform operations. It is suitable for almost all manufacturing processes including handling, assembling, welding, polishing, spraying, and dispensing.

According to the program input method, they are divided into programming input robot and teaching input robot. The former transfers the program files that have been compiled on the computer to the robot control cabinet through the RS232 serial port or Ethernet. This kind of industrial robot, which can be reprogrammed according to the changing needs of the working environment, works well in the flexible manufacturing process of small batches and multiple varieties with balanced and high efficiency, and makes an important part of the flexible manufacturing system (FMS).

An industrial robot that teaches input programs is called a teaching input robot. There are two teaching methods: one is that the operator uses a manual controller (teaching control box) to transmit the command signal to the drive system, so that the actuator can perform the required action sequence and motion trail; the other is that the operator directly leads the actuator and performs the required action sequence and movement trail. During the teaching process, the information of the working program is automatically stored in the program memory. When the robot automatically works, the control system detects the corresponding information from the program memory, and transmits the command signal to the drive mechanism to make the actuator perform the various teaching actions.

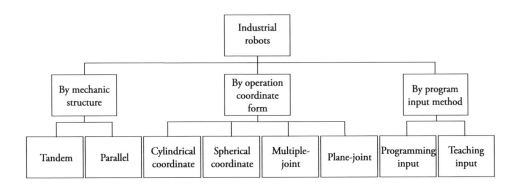

Certainly, the robot body alone cannot complete any tasks. Only after system integration can end customers use it. Therefore, in the manufacturing fields of injection molding, punching, polishing, spraying, fitting, welding, carving, die-casting, assembly, and loading and unloading, different system integration solutions are adopted, and welding robots, loading and unloading robots, spraying robots, assembly robots and so on are created to handles different industrial tasks.

Welding robots are industrial robots with welding tongs or welding (cutting) guns on their final shaft flange so that they can conduct welding, cutting or thermal spraying. This kind of robot has many merits, including stable and better welding quality, and the ability to reflect the welding quality in the form of numerical values; more reasonable labor intensity of workers; the ability to work in hazardous environments; and lower requirements for workers' operating skills. Loading and unloading robots can meet the requirements of fast/mass processing rhythm, of lower labor costs, and higher production efficiency, so it has become an ideal choice for more and more factories. The loading and unloading robot system has high efficiency and high stability, simple structure and easy maintenance, thus able to produce different kinds of products. For users, it can quickly adjust product structure and expand production capacity, and greatly reduce the labor intensity of industrial workers.

Spraying robots are also called paint spraying robots. They are industrial robots that can automatically spray paint or spray other coatings. Generally hydraulically driven, they are characterized by fast motion and good explosion-proof performance. Via hand-to-hand guidance or point-to-point display, they can teach. Painting robots are widely used in craft production departments such as automobiles, instruments, electrical appliances, and enamel.

Assembly robots are the core equipment of the flexible automated assembly system, which is composed of a robot manipulator, a controller, an end effector and a sensor system. They are used in the assembly of various electrical appliances, small motors, automobiles and their components, computers, toys, mechanical and electrical products, and their components.

Industrial robots are widely used in various industries such as electronic and electrical industry, automobile, rubber and plastic, food and beverage, chemical industry, foundry, and metallurgy. The system integration has a huge market space, with thousands of companies across China at present, and competition is getting increasingly fierce.

Electronic and electrical industry: in mobile phone production, the application of automated systems such as sorting and packing, film tearing systems, laser plastic welding, and high-speed four-axis palletizing robots, is suitable for a series of processes such as touch screen inspection, scrubbing, and film sticking. This type of robot is specially made according to the needs of the electronics production industry. Their miniaturization and simplification characteristics enable the high precision and high efficiency of electronic assembly, which meets the demand for increasingly refined electronic assembly of processing equipment, and greatly improves the efficiency of automated processing.

Automobile: In the production of automobile bodies, a large number of applications such as die-casting, welding, inspection, punching, and spraying need to be completed by industrial robots. In particular, their application in the automotive welding process has gained popularity, which has greatly improved the automation level of the workshops. More industrial robots will be installed in automobile forging workshops, punching workshops, engine workshops, painting workshops, etc.

Rubber and plastics: Injection molding machines and tools process plastic raw materials into fine and durable finished or semi-finished products. This process often requires the participation of industrial robots. They are not only suitable for operations under clean room environmental standards, but also can complete high-intensity operations next to the injection molding machines, so they can effectively improve the economic benefits of various processes. Industrial robots have the merits of being fast, efficient, flexible, sturdy and durable, and have a strong bearing capacity, which ensure the competitiveness of plastics manufacturers in the market.

Foundry: The environment for the foundry operation is rather poor, thus it is of great significance for industrial robots to replace human labor. With their modular structure design, flexible control system, and dedicated application software, the robots can meet the highest requirements of the entire automation application in the foundry industry. They are not only waterproof, but also resistant to dirt and heat. They can be used to take out the workpiece directly beside, inside and above the injection molding machine, and also reliably connect the process unit and the production unit.

3.4.3 International Layout of Industrial Robots

Zi Bin, Dean of the School of Mechanical Engineering, Hefei University of Technology, China, gave a speech at the CAIRDC2021 China Artificial Intelligence and Robot Developers Conference, "Manufacturing is the cornerstone of a country's industrial development, and robots are the crown jewel of manufacturing. Their research and development, manufacturing, and application are important indicators to a country's innovation capability and level of high-end manufacturing."

Since the birth of the world's first robot in the late 1950s, industrialized countries have established a complete industrial robot industry system, and mastered core technologies and product applications. There are four major families of industrial robots (ABB, Kuka, FANUC, and Yaskawa).

The United States: The Birthplace of Industrial Robots
The United States is the birthplace of industrial robots. With the development mode of integrated application, American engineering companies purchase and import industrial robot mainframes and complete sets of supporting equipment worldwide, and conduct integrated production line design, peripheral equipment R&D, and integrated debugging.

In 1954, George Devol, an American, patented industrial robots for the first time. In 1956, he and Joseph Engelberger founded the world's first robot company Unimation, and in 1959 they developed the world's first industrial robot, Unimate.

In 1960, AMF, founded by Harry Johnson and Veljko Milenkovic, made the world's first cylindrical coordinate industrial robot Verstran. In 1961, the Unimate 1900 series became the first mass-produced industrial robot, and was installed at a General Motors plant in New Jersey. It was mainly used in the manufacture of door and window handles, lighting equipment, gear lever handles and other vehicle-mounted hardware. In 1962, AMF installed six Verstran robots at a Ford plant in Canton, USA.

Unimate and Verstran are considered to be the earliest industrial robots in the world. They mark the official birth of robots. Since then, robot production research is mostly based on these two.

In the nearly 20 years since the birth of industrial robots, despite rapid technological innovation and advancement, they made little appearance in the market. In the 1960s, there were merely a handful of manufacturers and downstream customers of industrial robots. Till 1969, the global annual sales of industrial robots were only US$150 million. Because of the company's strategy, Unimation controlled nearly 80% of the U.S. industrial robot market at the time, but it didn't profit until 1975.

In the late 1970s, the political and industrial circles in the U.S. began to pay more attention to the application of robots, but their technical routes were more focused on special fields such as military, space, ocean, and nuclear engineering. This gave Japan, which paid more attention to the application of robots, the chance to catch up with the U.S. in the production and application of industrial robots, and to form a complete industrial chain.

Since the 1980s, the U.S. government began to formulate policies to stimulate the development of the industrial robot industry. On the one hand, it encouraged the research and application of robots in the industry, and on the other hand, it also granted more research funds to robots. The U.S. robot industry thus reached its second development climax. The number of robots soared from about 3,500 units in 1980 to 20,000 units in 1985 (the compound annual growth rate during the period was 41%). In the mid and late 1980s, with the maturity of the robot application technology of local American manufacturers, American robot manufacturers began to study and produce second-generation robots with perception systems such as vision and force, and soon occupied 60% of the American industrial robot market, thus returning to lead the industry.

At present, the development of industrial robots in the U.S. is rising at a steady pace. With the expanding application of technologies such as robot innovation research and human-machine collaboration development, the domestic sales of industrial robots in the U.S. reached 40,300 units in 2018, an increase of 21.6% year-on-year; in 2019, the number reached 33,300 units, a year-on-year decrease of 17.5%. The compound annual growth rate for the past decade was 17.16%.

Japan: Kingdom of Industrial Robots

Japan learned and perfected the industrial robot technology from the U.S., and earned itself the title "Kingdom of Industrial Robots." Japan has the most complete industrial chain of industrial robots, and the biggest industrial scale and strength in the world. There are internationally influential industrial robot suppliers in Japan such as FANUC, Yaskawa, and Kawasaki Heavy Industries, and core component suppliers such as Nabtesco, which is basically in a monopoly position in the world.

Japan adopts a development model based on the division of labor on a complete industrial chain of industrial robots, and continues to develop new industrial robots and mass produce industrial robot products; for the specific processes and demands of different industries, application engineering integration companies develop the integrated application for the complete system of industrial robot production lines. Japan has remained dominant in the industrial robot industry for a long time, holding a 60% global market share, which once exceeded 90%.

Multiple factors such as Japanese culture, automobile industry demand, insufficient labor, and preferential industrial policies have contributed to the development of the Japanese industrial robot industry.

For a long time, Japan has faced the problem of lack of labor, therefore it has a high demand for high-capacity and automated industrial robots, and pays more attention to the practical application technology of industrial robots. In 1967, after Kawasaki Heavy Industries Co., Ltd. took the lead in introducing industrial robot technology from the U.S., the government and enterprises prioritized the development of industrial robots. Since the 1970s, they have vigorously promoted the development and popularization of related technologies and applications. Next, Japan quickly surpassed the U.S. and ranked No.1 in the world in terms of industrial robots for a long time.

From the policy layer, the Japanese government has long been committed to promoting the development of the robotics industry by formulating a series of policies to support it since the 1970s. The Japanese government promulgated the "Mechanical and Electrical Law" and "Mechanical Industry Promotion Act" in 1971, and reorganized the former Industrial Robot Association into the Japan Industrial Robot Association (JIRA) to promote the development of robot manufacturing; in 1978, the "Machine Information Law" came out; In 1980, the "Financial Investment, Financing, and Lease System" and the "Small and Medium-sized Enterprise Equipment Modernization Loan System and Equipment Debt and Credit System" were promulgated to popularize robots to SMEs in the form of leasing; in 1984, there was "FMS Machine Leasing System" and the "Mechanical and Electrical Integration Tax System"; in 1985, the "High-Tech Tax System," "The Tax System for Promoting Infrastructure Development," and "On Strengthening the Technical Basic Tax System for Small and Medium-sized Enterprises" were released to promote preferential tax reductions and encourage the development of leading technologies, and an international robot FA technology center was established; in 1991, the Ministry of International Trade and Industry initiated a large R&D project of micro-machine technology.

Recently, the development of robots in Japan continues, and the government's support in this field remains as strong as before, and is predicted to get stronger. Under the circumstance where the demand in the international market dropped due to the global economic downturn, Japanese robot production in 2019 fell 19.5% year-on-year to 173,477 units, and sales down by 18.35% year-on-year to 175,702 units. However, as the market recovers, the production in the first half of 2020 increased by 6% year-on-year to 88,856 units, and sales also went up by 6.62% year-on-year to 91,298 units.

Europe: Germany, the Protagonist

Europe is a major part of industrial robots in the world. It has achieved complete autonomy of core components such as sensors, controllers, and precision reducers. KUKA of Germany, ABB of Switzerland, COMAU of Italy, and AutoTech Robotics of the UK are the world's top industrial robot manufacturers. Europe adopts the mode of providing users with system integration solutions, while the industrial robot manufacturers undertake and complete the system design and integration debugging of the production and application process of the industrial robots.

Germany has the largest market in Europe. In addition, German robots are in the lead in the fields of human-computer interaction, machine vision, and machine interconnection.

From the perspective of the development of German robots, Germany introduced industrial robots five or six years later than the UK and Sweden. In 1971, there were fewer than 50 robots. By 1972, there were no factories to manufacture them. In order to promote their development and application, the German government issued the "Working Conditions Improvement Plan" in the 1970s, stipulating that robots must replace humans to handle dangerous, toxic, and harmful jobs. Robots are not only able to greatly reduce production costs, but also improve manufacturing accuracy and quality, thereby building a positive and strong image of made in Germany.

The first robot automatic welding production line, which was created in 1971, was used to process the side panels of Daimler-Benz cars, employing the five-axis robots from Unimation of the U.S. In view of the automotive industry's demand for robots with highly reliable performance, KUKA, a German company, developed the first KUKA industrial robot in 1973. Since the 1980s, Germany has employed a large number of industrial robots in automobiles and electronics.

In 2004, the German government and the German Bundeslands entered into the "Research and Innovation Agreement," requiring the four major domestic research associations (i.e. The Max Planck Society for the Advancement of Science, The Helmholtz Association of German Research Centres, The Fraunhofer Society, and The Leibniz Federation of Science) should maintain their R&D expenditure at an annual growth rate of at least 3%. This has cultivated a large number of talents for the robotics industry and promoted the sustainable development of the robotics technology.

In 2010, the German government issued the "Germany 2020 High-Tech Strategy," which laid out a strategic plan for the robotics industry, and implemented the Industry 4.0 strategy in 2013, taking the lead in dividing the industry into four stages, thus pointing out the trend of intelligence. The physical entity of this trend is robots, machinery and equipment, and human-machine collaboration improve the intelligence of the production processes.

After years of development, SEW, FLENDER, and other German companies have become world-renowned reducer brands. Among them, KUKA's industrial robot applications in the automotive field have long been top of the global market, and there are other well-known German robot integration companies, such as REIS and DURR. In 2019, the sales of industrial robots in the German market reached 20,500 units, accounting for 4.85% of the global market, making it the fifth-largest market in the world.

South Korea: Towards the Climax in the 21st Century
South Korea's industrial robots started slightly later than the U.S., Japan, and Germany. It mainly developed industrial robots through the introduction of FANUC technology from Japan. As a result, Hyundai Heavy Industries and Samsung rose.

In 2010, the sales volume of industrial robots in South Korea surpassed Japan and became the world's number one, and its industrial robot usage density has stayed number one in the world. South Korea plans to supply no less than 700,000 robots in the five-year period from 2019 to 2023.

Since Hyundai Heavy Industries introduced technology from Japan's FANUC, South Korea has begun to develop industrial robots. It first introduced welding robots in 1978, which were mainly used for the automobile manufacturing. Since then, the industry and academia have started spontaneous technical research without government support. It was not until the end of the 1980s that the Korean government began to launch preferential R&D policies to support the development of the robotics industry. However, due to the impact of the Asian financial crisis in 1997, government funding and R&D almost stagnated in the next few years.

In 2002, as intelligent robots were created, the Ministry of Trade, Industry and Energy (MOTIE), Ministry of Science and ICT (MSIT) and other departments resumed their support for the robotics industry, increasing relevant funding and formulating plans.

In August 2003, MOTIE included intelligent service robots as one of the ten key industries for development, and made more efforts to cultivate talents related to the robotics industry. At this time, South Korea remained focused on domestic robots and personal robots. During the six years from 2002 to 2007, the South Korean government invested massively in technology development and market expansion, with a total investment of about 486.5 billion won, of which about 420.2 billion won was used for R&D, and 9.5 billion won for boosting market demand. It funded as many as 1259 projects in total.

In 2008, the Korean government promulgated the "Intelligent Robot Development and Popularization Law," which listed robots as a national strategic industry from the legal level, and formulated a basic plan for their development.

In 2012, the Ministry of Knowledge Economy (MKE) of South Korea issued the "Robot Future Strategy 2022," planning to invest 350 billion won in the development of the robotics industry. It aims at making the Korean robotic industry one of the top three in the world. In this strategy, industrial robots are the main development direction – the policies target the intelligent upgrade of industrial robots so that they become a pillar industry and organically integrate with other industries.

To implement the "Robot Future Strategy 2022," in 2013, MKE launched the "Second Intelligent Robot Action Plan (2014–2018)" and detailed the development goal of robot output value, export value, and global market share in the next 5 years. As of 2016, South Korean industrial robot manufacturers accounted for 5% of the global share.

With preferential policies, the Korean industrial robotic industry entered a period of rapid growth after 2000. From 2001 to 2011, the average annual growth rate of the total installed robots in South Korea reached 11.7%. The usage density of Korean industrial robots continues to rise, and so does the self-sufficiency rate of industrial robots. In 2019, South Korea's domestic robot sales reached 27,900 units, and the compound annual growth rate over the past ten years reached 13.53%.

China: Booming, Bigger Opportunities Than Challenges

The development of industrial robots in China is currently in full swing. Chinese industrial robots have entered a period of rapid growth after 2010, showing slower growth but higher sales and a 3–4 year cyclical characteristic, which is similar to the data of Japanese machinery orders. At present, Chinese industrial robots exhibit a clear trend of rebounding. They have entered a new round of economic recovery, which is expected to last until 2022 to 2023.

The "Robotics Industry Operation Situation from January to December 2020" issued by the Ministry of Industry and Information Technology shows that in 2020, a total of 237,000 units of industrial robots were produced, with a 19.1% year-on-year increase, setting a record for the highest annual output of industrial robots in China. And according to the National Bureau of Statistics, from January to February 2021, the output of industrial robots of industrial enterprises above designated size nationwide was 45,400 units, a 117.6% year-on-year increase, setting a new record for the same period in the previous years. From made in China to intelligently made in China, robots are playing an increasingly important role in this upgrade.

The development of industrial robots is basically owing to the scale effect of technological progress that lowers their price. In the initial stage of their development, high prices were once the main factor hindering many small and medium-sized enterprises from purchasing equipment and setting up intelligent production lines. With the market impact from domestic industrial robots, the advancement of manufacturing technology and the rapid decline of manufacturing costs, the price of industrial robots has exhibited a clear downward trend in recent years.

Take the Chinese import and export prices and global prices as examples. The average price of imported robots in China dropped from US$30,000/unit in 2009 to US$16,800/unit in 2016, with a gross annual growth rate of –7.9%; The export price also decreased from US$29,300/unit in 2011 to US$5,200/unit in 2016, with a gross annual growth rate of –29.3%; the global average price went down from US$63,300/unit in 2009 to US$44,500/unit in 2016, with a gross annual growth rate of –4.9%.

In addition, in the manufacturing industry, higher and higher human labor cost has dealt a huge blow to labor-intensive industries. Enterprises are turning to more economical production modes to reduce costs, and it is inevitable that machines replace workers. As the Chinese economy develops rapidly, the average annual salary of employees in the manufacturing industry has soared from RMB26,800 in 2009 to RMB64,500 in 2017, with a gross annual growth rate of 11.59%.

Meanwhile, the shrinking working-age population and the disappearing demographic dividend are forcing the industry to develop. From the perspective of population structure, the proportion of the population aged 15–64 in China began to decline from a high of 74.50% in 2010 to only 71.82% in 2017. And the natural population growth rate has been stable at the low level of around 5% in the past 15 years, which means that the proportion of the working-age population in China will stay at a relatively low level in the future, thus creating an urgent demand for the development of industrial automation.

Obviously, preferential policies have contributed greatly to the rapid development of industrial robots. The earliest development of industrial robots in China can be traced back to Project 863 of the Ministry of Science and Technology, which supported the R&D of robotics-related technologies. In February 2006, the State Council issued the "Outline of the National Medium- and Long-term Science and Technology

Development Plan (2006–2020)." For the first time, intelligent robots were included in the advanced manufacturing technology of cutting-edge technologies. Next, as "Made in China 2025" strategy was made and implemented in 2015, local governments at all levels have actively promoted the implementation of regional planning policies, and the Chinese industrial robot industry has also achieved rapid development.

It is an important trend that robots replace humans in the manufacturing industry. It is the basis of intelligent manufacturing and the guarantee of industrial automation, digitization, and intelligentization in the future. Although the application of Chinese industrial robots in manufacturing and industrial facilities is widened quickly, there is still a big gap between Chinese industrial robots and those of developed countries in terms of manufacturing and application.

From the perspective of manufacturing, industrial robots follow the development path of taking automation as the underlying technology, and moving toward digitization, networking, and intelligentization. The higher they go, the more help they need from chips, software, and algorithms. Even though China's manufacturing output value has surpassed that of the U.S. since 2010, and the scale of development is large, the quality and efficiency are not high enough, meaning still a lot of room for improvement.

Today, industrial robot manufacturing in developed countries has reached the intelligent stage, while China is still a rookie. According to China Business News reports, domestic industrial robots have occupied most of the domestic market in many segments by virtue of their advantages in cost performance and channels. However, they are still behind international advanced levels in terms of key technologies, materials, and components.

Among the newly installed robots, 71% of the components are from abroad, making the localization rate less than 30%. Among the components, the localization rate of the three most important upstream components – reducer, servo motor and controller is 30%, 22%, and 35% respectively, which are low numbers. In terms of product accuracy and stability, there is still much space for improvement.

Also, there is serious homogenization in Chinese industrial robots. At present, the products of many industrial robot brands are mimicking each other in performance, appearance, technology, and marketing. Even the core technology and production goals of the products that different manufacturers develop are the same. Therefore, in the current industrial robot market, there are rare outstanding and competitive robot products.

In terms of application, industrial robots in developed countries have already installed a complete set of equipment in industrial production lines, and they have been skilled in the use of industrial robots. The industrial robots can perform functions without the need to be equipped with specialized operators. Instead, Chinese industrial robot application still requires the presence of specialized industrial robot operators to assist in the operation.

Industrial robots were popularized in Japan in the 1920s, but so far, the development of Chinese industrial robots is still at an early stage, and it is facing major challenges such as shifting to a high-end level, undertaking international advanced manufacturing, and international division of labor.

The continuous development and innovation of industrial robots make higher requirements for practitioners, and there are increasingly tense contradictions between the supply and demand of talents in this field. In manufacturing, servo motors, controllers, and reducers have become the main bottlenecks restricting the industrial robot industry in China. There is a critical shortage of technical talents in these areas. In application, there is a huge gap in the talents to apply industrial robots such as corresponding operation and maintenance, system installation and commissioning, and system integration.

The greatest contribution of industrial robots lies in the improvement of manufacturing productivity rather than that of the robots themselves. With the advancement of control, drive and sensing technology, the scope of work that robots can take on keeps expanding, and the changes it brings will be far-reaching and beyond imagination.

3.5 Information Technology: Energizing Traditional Manufacturing

The information industry is the basic, leading, and strategic industry of the national economy. The boom of information technology has reshaped the world. It has strongly driven economic growth and boosted industrial upgrading, occupying an important strategic position in the integration of informationization and industrialization. Whether it is the smart industry represented by collaborative manufacturing and 3D printing, or the advanced manufacturing characterized by smart design, smart manufacturing, smart operation, smart management, and smart products, they are inseparable from the strategic support from the information industry.

The integration of informationization and industrialization has generated the 1 + 1 ≥ 2 economic effect, and information technology has brought advanced technologies, concepts, and management modes to the manufacturing industry. The integration of information technology and the various links involved in manufacturing will be conducive to transforming and upgrading the manufacturing industry in terms of product design, equipment, management, and marketing, to rationally utilizing labor, technology, and resources, to reducing resource waste, and to changing the original consumer demand and product structure. Through the influence of informatization on each link, the overall technical efficiency of the manufacturing industry is improved, thus moving the actual output closer to the production possibility curve and promoting the informatization of the manufacturing processes.

3.5.1 Cluster Information Technology

Information Technology (IT) is a collection of various technologies mainly used to manage and process information. IT mainly applies computer science and communication technology to design, develop, install and implement information systems and application software, including industrial software, cloud computing, Internet of Things, and the Internet.

Industrial Software

Industrial software is dedicated or mainly used in industrial manufacturing to provide technical and knowledge support for product R&D, production and operation management, supply chain coordination, and equipment and product intelligence. It exists in all elements and links in the industrial field, and is closely integrated with business processes, industrial products, and industrial equipment. Comprehensively, as the fusion agent of informatization and industrialization, it supports various industrial activities such as R&D, design, manufacturing, and operation management. As an entry point, a breakthrough and an important starting point for the integration of informatization and industrialization, it is of great significance to promote the transformation and upgrading of China's industry and maintain stable and rapid economic growth.

The significance of industrial software is mostly reflected in the digital support for the product R&D, refined management and decision support for business operation, automated control and digital manufacturing of the manufacturing processes, intelligent value enhancement of device-level embedded chips and software, and the integration and collaboration of internal and external enterprises and upstream and downstream industries. Industrial software can increase product value, improve labor productivity, provide refined management, reduce production costs, and enhance the core competitiveness of enterprises, thus becoming the core of modern industrial equipment and products.

Industrial software plays a paramount role in the integration of industrialization and industrialization. It is the embodiment of industrial soft advantages. By enabling mechanized, electrified, and automated production equipment in the sense of traditional industrialization to have the core technologies of digitization, intelligentization, and networking, it helps enterprises build a networked, collaborative, and open product design and manufacturing platform oriented at the full life cycle of products, so that they can form the soft advantage of informatization on the basis of the hard advantage of industrialization.

Cloud Computing

The concept of cloud is applied to the Internet to describe the form and essence of the future Internet. It refers to the global computer hardware and software, including servers, terminal equipment, and network data lines. It integrates and connects quickly and organically according to the user needs, so that they can enjoy the cloud service at super low cost or free.

Sun CEO Scott McNealy is the first ever documented to have proposed the concept of cloud computing. He proposed the concept of "network computer" in the 1990s, suggesting that the network is ubiquitous. For over the following 20 years, Sun's technical team, IT and Internet industries have been exploring and practicing to provide users with an open infrastructure service platform with lower cost, easier operation, and more secure data.

On May 21, 2010, at the 2nd China Cloud Computing Conference, Hongmeng Group Chairman Mr. Zheng Shibao delivered a speech "Viewing Cloud Computing from Life, Holism vs. Reductionism," integrating cloud computing into Eastern science and philosophy. With Chinese intelligent thinking, he defined it from the viewpoints of holism and system theory: cloud computing targets an application, connects a large amount of necessary hardware and software according to a certain structural system through the Internet, and builds a virtual resource service center with the smallest internal consumption and the highest efficiency by constantly adjusting the structure system as the needs change.

In short, cloud computing connects the tangible and intangible resources associated with the Internet to form a platform where users can do what they want in accordance with the rules. This also means that computing will more become a service. Through the Internet, a large amount of computing power from afar will be used locally.

Documents, emails, and other data will be stored online, or more precisely, "stored on the cloud."

The new approach makes a huge promise. For enterprises, by switching to cloud-based email, accounting and customer tracking systems, they can reduce complexity and maintenance costs, because everything runs in a web browser. Enterprises are no longer isolated. On the cloud computing network platform, the information body is displayed freely and equally on the open cloud platform as the search engine target core optimization technology amplifies the valuable information of the operators, so that customers can quickly find the enterprises through the Internet. At the same time, cloud computing service providers can also profit through economies of scale.

Moreover, the cloud computing network platform adopts technical means to perform community management and division of social public information resources, breaking the situation where a minority of interest groups exclusively enjoy public information resources. As a huge community website operator owns the public information resources of the community, it can convert these information resources into economic benefits.

The Internet of Things

From the PC Internet to the mobile Internet, and to the Internet of Things (IoT), all rounds of information revolution point to the same keyword – "connection." If the Internet is considered to have bridged the connection between "people and people" and "people and information," the IoT goes one step further and comprehensively connects "people and things" and "things and things." Certainly, the development of the IoT has also experienced a long period of introduction, accumulation, and verification.

In 2008, the first International Internet of Things Conference was held, where the number of IoT devices exceeded the number of attendants for the first time. In the introduction stage of the IoT, it was characterized by the introduction of related concepts of the IoT and the connection of early IoT devices. In 2013, Google Glass was released, revolutionizing the IoT and wearable device technology. In 2016, various elements that shape the ecosystem of the IoT industry were ready. The precipitation period of the IoT mainly involved the trial and error and precipitation of some sensing and communication technologies.

As soon as the various elements of the IoT industry chain were basically in place, the scale effect of the IoT on the transformation of the national economy and industry was initially showed. 2018 to 2019 was the opening period for the market to verify the implementation of IoT technology solutions. During the IoT verification period, the drive from technology, policies, and industry giants still plays a significant role in the development of the IoT industry. However, it cannot be ignored that the influence of market demand factors is expanding.

Since Kevin Ashton proposed the term "Internet of Things" in 1999, it has gradually evolved from an embryonic form to a new engine that drives global economic growth. The new wave of technology has knocked open the door to a new age, and set a unique tone for it.

Although from the perspective of the connected objects, the IoT merely adds a variety of "things," it has an extremely far-reaching impact on the expansion and sublimation of the connotation of the connection. The IoT is no longer a single connection center of "people." Things and things can be connected independently without human manipulation, which ensures the objectivity, real-timeness and comprehensiveness of the content to be delivered via the connection to a certain extent. In addition, the IoT connects every detail of the real world to the network, creating a system where virtuality (information, data, process) and reality (human, machine, and commodity) reflect and connect one another. The physical entities establish their own digital twin in the virtual world, making the state of itself traceable, analyzable and predictable.

In the environment of the IoT, on the one hand, everything is an entrance. Except for the data generated from the users' active interaction, many passive user data will be recorded in real time and without being noticed. Therefore, enterprises can understand user needs comprehensively, three-dimensionally, and dynamically. On the other hand, smart factories in the age of the IoT can quickly meet users' continuously iterative needs for customization through flexible production lines and transparent supply chains.

Compared with the approximately 5 billion device accesses of the mobile Internet, the scale of IoT connections will be at least an order of magnitude bigger, covering everything from wearable devices, smart homes, autopilot vehicles, connected factories, and smart cities. In the future IoT age, the devices connected to the network will be more intelligent and data applications richer without being limited to their current simple item status and location information. This new wave that the IoT leads is going to fundamentally change the way of life we are used to, and reshape the pattern of the global industrial economy.

The Mobile Internet

Relying on the development of electronic information technology, the mobile Internet combines network technology and mobile communication technology, while wireless communication technology can obtain various network information with the help of the intelligence of the client terminals. Therefore, the mobile Internet is also regarded as a new business model, which involves various contents of applications, software, and terminals.

Mobile Internet is the road for the development of IT and its industry. It constantly creates unconventional new business models, new markets, new rules, and new concepts, quietly changes the configuration relationship between the core elements of the IT system, and cultivates a new generation of information technology and related industries, such as cloud computing, big data, and IoT. It is an important area in which groundbreaking innovations are taking place and about to take place in information technology.

Mobile Internet has undergone the budding period where 2G and WAP were the main applications, the cultivation period where 3D networks and smartphones were the leading roles, and the high-speed growth period. It has now entered a comprehensive development period with 4G network construction as the mainstay and 5G network as the traction. The emerging industrial innovations in mobile shopping, mobile games, mobile advertising, mobile payment, mobile search, mobile medical care, industrial Internet, and other mobile Internet platform services, and information services will push the mobile Internet industry to the stage of deepened development of application and service.

The fifth generation of mobile communications (5G) has become the focus of the present and future global industry. It is going to lead the mobile Internet into a new age. 5G is a brand-new and groundbreaking starting point. It will meet the global expectations for the entire industry upgrading. 5G is not only a revolutionary step forward for the communications industry, but will also create unprecedented business opportunities for all walks of life.

5G builds the core basic capabilities to connect everything. This not only brings faster and better network communications, but also shoulders the historical mission of empowering all walks of life. The 5G ultra-high speed and geometrically increased connection density are the basic guarantees for the Internet of Everything. They make it possible for the application of the Industrial Internet to cover the entire industrial chain and all the production processes. In the chemical, machinery, electric power industries, many enterprises have relied on 5G to enable the industrial Internet to perform supply chain management, real-time remote control of the production process, equipment collaboration, flexible manufacturing, inventory management, and delivery management. As a result, the configuration efficiency, production efficiency and product quality have been significantly improved, while the operation and management costs for the enterprises have been effectively lowered.

5G opens a new chapter in the development of the Internet and writes it a new future. In addition, 5G, as the leader of "new infrastructure," is the infrastructure of other "new infrastructure" such as artificial intelligence and big data centers. In fact, the present digital economy development has stepped into a stage of cross-industry integration, 5G development is accelerating, and new industries, new business modes, and new models are constantly springing up.

With the development of cloud computing, the IoT and the mobile Internet, and the deepened integration of the three networks, data services have become the broadband mainstream. The remoteization of computing resources has broken the closed walls of the original communication industry. Under the dual drive of technology and the market, the information industry and the communication industry are merging, laying the foundation for a new wave of technological revolution.

In the age of smart networks, more information technology has been given the meaning of informationization and intelligence. New Internet services, cloud computing, and various commercial applications have become the main content of information; sensor networks, smart terminals, and all-IP networks lead to the new communication mode represented by cloud management terminal. These future industries themselves become strategic new industries. They serve as the leaders of China's economic growth in the future, while providing informational means for other new industrialized industries. Cloud computing, the IoT, industry informatization, intelligent integrated terminals, and mobile Internet will become the key engines to promote the deep integration of the informatization and industrialization in China.

The information industry realizes digitization, informatization, and intelligentization by integrating itself with the new generation of information technology and accelerates the growth of other strategic new industries. The integrated development and application of the IoT, cloud computing, big data, mobile Internet, and next-generation networks and industries have enabled the construction of smart cities. The higher the degree of integration of the IoT with various technologies and industries, the wider the scope of integration, the higher the degree of intelligence of the IoT, and the higher the degree of urban intelligence. As various networks continue to advance and people demand information processing anytime and anywhere, the integration of three networks will evolve to the integration of N (4, 5, 6 . . .) networks, and eventually become a unified intelligent IoT after a long development.

3.5.2   Deepen the Development of Manufacturing

Information technology (IT) has had a profound impact on the development of human society. It is transforming the traditional manufacturing, setting off a new intelligent technological revolution in industrial manufacturing.

Primarily, IT can improve the technical level and added value of products, and propel product upgrades and updates. First, it is integrated into every link of product manufacturing, development and design. Adopting IT design in the process of research and design can enable comprehensive management from the design process to design data, thereby improving the innovation capability of manufactured products, saving cost, and accelerating the renewal cycle of manufactured products.

Second, the application of IT and tools can improve the design capabilities of manufacturers. At present, product R&D and design are getting increasingly complex, and higher and higher requirements are made for an enterprise's design capabilities. Only R&D and design based on IT can survive in the fierce market competition.

Third, IT software can improve the efficiency of R&D and design. Digital tools are employed such as computer-aided design and computer-aided manufacturing to improve the efficiency of research and design units, while product data management is applied to improve that of research and design organizations.

Next, regarding equipment transformation, IT is integrated into the manufacturing processes, combining with machine tools, to achieve excellent automation and intelligence, establish an IT workshop production line, and make production equipment smarter. The earliest application of IT in the manufacturing industry is its equipment foundation. The organic combination of manufacturing equipment and IT improves the production efficiency of the manufacturing industry and transforms the traditional manufacturing industry into intelligent, flexible, and precise new manufacturing.

In this process, the combination of the IoT, cloud technology, three-networks-in-one and the traditional enterprise informatization will jointly promote the establishment of a new type of industrialization and thoroughly transforms the industrial system. For example, IT will transform the industry from dependent on resources and investment to dependent on technological progress, especially the integration and application of IT; from the traditional extensive quantitative expansion to the qualitative transformation of technological capabilities, which is embodied in advanced manufacturing and smart products and equipment. As a result, there will be a breakthrough in the core link of the manufacturing value chain, and a shift to the high-end. In these transitions, software and information service industry, new-generation IT, chip design, and embedded systems all play a pivotal role.

Then, IT can improve the management level of manufacturers. In this regard, on the one hand, through IT, innovations in enterprise management models can be achieved, including innovation in customer relationship management (CRM), innovation in modern enterprise management structures, and governance models. On the other hand, IT can also make enterprises better at information analysis and decision-making capabilities.

At present, the manufacturing industry has adopted enterprise resource planning (ERP) to help business managers understand the business situation more accurately, respond quickly to market demand, and provide decision-makers with timely data, so that manufacturing companies can better face the fierce market competition and realize their IT.

Finally, IT will also promote innovation in the marketing system. For example, the application of e-commerce enhances the efficiency of corporate marketing, promotes the innovation of the marketing model of manufacturers, saves transaction costs, reduces intermediate complex links, and renders customers with more convenient and efficient services. Therefore, the corresponding speed of the market and the speed of response from customer service are faster, thus innovating the marketing system of manufacturers.

The informatized and integrated development is inevitable. Integration is the trend of global economic development. Under the circumstances of economic globalization and the development of economic and social information, only by using modern IT and the integrated business that takes IT as a means to promote the transformation and upgrading of traditional industries, and to integrate them with emerging industries, can the overall economic development and upgrading of the entire society be fostered.

CHAPTER 4

Manufacturing Mode Group

4.1 Lean Production: Less but Better

Over 20 years ago, James P. Womack, Daniel T. Jones, and Daniel Roos first systematically elaborated Lean Production (LP) in their book *The Machine That Changed the World*. LP was in fact the Just in Time (JIT) production mode of Toyota, which won acclaim from several leading experts of the International Motorized Vehicle Project (IMVP) of the Massachusetts Institute of Technology (MIT). Thereafter, as an advanced manufacturing mode, it has been known to the world and continuously studied.

Lean, that is, less but better. There is no input of extra production factors, but the necessary quantity of products urgently needed in the market (or products urgently needed in the next process) are produced at the proper time; Lean also means all business activities must be profitable, effective, and economical.

LP produces high-quality products by effectively eliminating waste, unreasonable, and non-value-added links in production. As a production organization system and method that is widely accepted in the industry, it has been introduced by an increasing number of enterprises so that they maintain a low-cost and high-quality competitive advantage in the fierce global competition.

4.1.1   LP Building Structure

The manufacturing mode of LP is the product of the post-war market restriction of "resource scarcity" and "multiple varieties, small quantity" upon the Japanese automobile industry. Starting from Sakichi Toyoda, it was not until the 1960s that the LP model was gradually perfected through the joint efforts of Kiichiro Toyoda and Ohno Taiichi. In short, LP is a production method that aims to minimize the resources occupied by the enterprise's production and reduce management and operating costs thereof.

In recent years, LP and the soaring Japanese automobile industry have attracted great attention of scholars and engineers from various countries, who began to study and explore the mechanism and structure of LP. Among them, Wang Pin of the Tianjin University of Science and Technology, having visited 28 companies including Toyota Motor Company and the cooperative enterprises thereof, proposed a new LP building structure: JIT, flexible automation production (FAP), total quality management (TQM) and specialized cooperation production (SCP). The structure foundation is concurrent engineering (CE) and team work (TW) that are based on computer networks.

JIT is the origin and core of LP. It aims to only produce the necessary quantity of necessary products when necessary, eliminate all waste and make no redundant finished products. Kanban management is the implementation of specific measures for JIT. It is an information system that controls the production activities of each process during the production. It adopts the "fetching system," that is, the subsequent process produces according to market needs. When there is a shortage of work in process in this method, it takes the same amount of work in process from the previous process, so as to build the whole-process pulling control system without producing one redundant product. In this way, the kanban is circulated between the various processes in the production, so that the information related to the time, quantity, and variety of the taking and production processes is transmitted from the downstream to the upstream of the production, and the relatively independent processes are connected as an organic whole.

Kanban management can prevent overproduction, thereby completely eliminating the waste of product volume and all kinds of indirect waste derived from it. It also fully exposes the reasons for defective products, hidden problems, and unreasonable links in the production process. By thoroughly addressing the problems, Kanban management removes all kinds of waste that lead to higher cost, and make the production process rational, efficient, and flexible.

Contrary to rigid automation that is scattered, has a fixed pace, and adopts flow production, LP adopts moderate FAP, is based on group technology (GT), and employs the automated production system composed of numerically-controlled machine tools, machining centers or flexible production systems, robotics, and automated testing technology. It is a critical means for modern production to adapt to the requirements of a rich variety, fine quality, and technical efficiency.

Toyota applies flexible automation. In the event of an abnormality on the production line, each station is equipped with a rope pulling device that can stop the line at any time, and the photoelectric display board suspended in an eye-catching position synchronously displays relevant information. So, immediate measures can be taken to resolve the problem. In addition, as reasonable suggestions are taken, many protection devices and facilities that embody humanistic concepts such as labor-saving seats, foolproof warning devices, and minute exchanges of the die were installed on the production site to prevent operating errors.

Total Quality Management (TQM) was renamed from Total Quality Control (TQC). TQM aims to mobilize the enthusiasm of every employee, go deep into every production link, ensure product quality, and nip product quality problems in the bud. With this in mind, the TQM teams of the Japanese enterprises have also incorporated improvements of the process plan, the reduction of consumption, and the enhancement of efficiency into quality management. Compared with ISO9001, TQM has higher and stricter requirements.

Specialized cooperation production develops as the division of labor in society develops. Professional division of labor splits social production into many independent specialized production units while cooperation unites each unit into an organic whole.

Specialized production is conducive to making use of the latest scientific and technological achievements so that the level of production mechanization and automation is higher; to technical training of workers so that their production technological operations level is improved; to enhancing the utilization rate of production hours and production equipment, and product quality; to exploiting production potential so that human and material resources are fully and effectively used, thus leading to better economic results; to shortening the time of plant construction and saving capital construction investment; to simplifying production management and improving the level of corporate management; to changing the "big and complete" or "small but complete" production structure of enterprises, and to promoting the development of production. Specialized cooperation is the inevitable future of production development. In the mass production of modern society, only by continuously improving the level of professional cooperation can enterprises survive and develop.

Computer network-based concurrent engineering (CE) and team work (TW) are the basis of Toyota's LP. CE is the integrated and parallel design of products and related processes (including manufacturing processes and support processes). Product developers are required to consider all factors throughout the product's entire life cycle from the concept to scrap at the beginning of the design. And Toyota Motor considers user needs, quality, cost, schedule, environmental friendliness, service life, and recycling of scrapped materials at the beginning of the design.

4.1.2   Revolution of Production Modes

LP reduces costs, shortens the production cycle and improves quality by eliminating non-value-added activities in all enterprise links, and eventually overcomes the challenges of small orders, multiple varieties and personalized demand in the ever-changing market. It is precisely because LP shakes the basic principles and foundation of mass production that it has revolutionized production modes.

James Womack and his colleagues concluded in *The Machine That Changed the World* that Toyota had indeed revolutionized manufacturing; the factories that conducted old mass production modes could not compete with it. The new best, LP could be successfully transplanted to the new environment; when LP is adopted and inevitably applied beyond the automotive industry, it would change almost everything in all industries, including consumer choices, the nature of work, a company's wealth, and ultimately the future of a nation.

First, LP has changed the separation of the production process, and proposed the idea of production management based on the process. In the handicraft stage, the production process was closely connected. In the mass production stage, people have introduced work-in-process inventory and cost inventory so that sales, production, and different production processes can be separated from each other, and economies of scale be taken advantage of to increase production efficiency. Since the development of mass production, the market has changed. There is oversupply and diversified customer demand, meaning that the economies of scale of mass production are no longer fit for the times. Therefore, LP shrinks the production volume to a single piece and frees the inventory in each link via rapid adjustment. Meanwhile, when the management of the entire production chain is considered based on the value-added process, all links of the production system can be totally connected.

Secondly, in terms of organizational management, LP has groundbreakingly played the role of team work, changed the pyramid-shaped organizational system, and clarified the trend of flat organizational development. James P. Womack pinpointed that the dynamic working group is the core of the lean factory operation. Regarding factory field management, quality management, and product development, LP focuses on team-based work and management. In addition, it redistributes responsibilities and powers in production, and clearly proposes the guiding principles for authorization. It grants most of the power in production to front-line workers, and proposes that production decisions be made at the lowest possible level in the organizational structure. In a way, the flat hierarchical structure, wide control range, multi-functional team and authorized employees make up the organizational innovative characteristics of LP.

At last, in work process design and knowledge management, LP is characterized by frequent work rotation, appropriate task scope, and minimum number of work classifications. The true core of the Toyota Production System lies in the professionalization and clarification of work content; the clarification of scientific working methods; and the simplification, directness, and automatic adjustment of various supply relationships, product, and service connections.

The key to LP is the management process, including the optimization of personnel organization and management, vigorous simplification of middle management, flattening of the organization, fewer indirect production personnel; promotion of production equalization and synchronization to realize zero inventory and flexible production; the application of a quality assurance system of the production process (including the entire supply chain) to achieve zero defects; the reduction of waste in any link until there is nothing; and finally the realization of a pull-type JIT production mode. LP is characterized by zero waste, excellence, and continuous improvement. Thus, simplification is its very core.

4.2　Green Manufacturing: Towards Environmental Friendliness

The development of the processing and manufacturing industry has empowered mankind to transform nature and obtain resources. People consume the products it produces either directly or indirectly, which greatly improves their living standards. As a pillar industry that creates wealth, over the past 100 years, when the processing and manufacturing industry has brought unprecedented civilization and wealth to mankind, it has also caused serious environmental problems.

At present, the deteriorating environmental pollution has raised unprecedented awareness of environmental protection, and the decoupling of global economic growth and the increase of carbon emissions has become inevitable. In China, the construction of ecological civilization is regarded as an important part of promoting the overall layout of the "five-sphere integrated plan" and the strategic layout of the "four-pronged comprehensive strategy." With the dual transformation of social digitalization and energy upgrading, green manufacturing is gaining more weight.

4.2.1 The Inevitable Green Manufacturing

In 1996, the American Society of Manufacturing Engineers (SME) released the Green Manufacturing bluebook titled *Green Manufacturing*. From then on, studies on this topic have sprung up all over the world. Furthermore, as environmental problems deteriorate and the society becomes digitalized, green manufacturing has become inevitable.

Green manufacturing, also known as environmentally conscious manufacturing or environment-oriented manufacturing is a modern manufacturing mode that comprehensively considers the negative impact on the environment and resource utilization. It aims to minimize the impact on the environment and optimize the utilization of resources throughout the life cycle of the product including design, manufacturing, packaging, transportation, use, scrap, and disposal. It has a broad connotation. In addition to protecting the environment and making the best use of limited resources, it also includes two levels of entire process control.

On the one hand, it is the process of making full use of resources, reducing environmental pollution, and realizing specific green manufacturing goals in the collective manufacturing process, that is, the material conversion; on the other hand, it refers to the broadest green manufacturing process that maximizes the utilization of resources and reduces environmental pollution by thoroughly considering resource and environmental issues in every link of the product lifecycle, including conception, design, manufacturing, assembly, transportation, sales, after-sales, scrap, and recycling.

At present, human society is faced with three major challenges of a worsening environment, limited resources, and shrinking population. The deteriorating and rapidly increasing environmental problems are posing a serious threat to the survival and development of the human society. Also, these problems are not isolated but intrinsically linked to the two major problems of resources and population. The resource problem is not only how to use the limited resources on this planet, but also the root of environmental problems.

Under this circumstance, the best use of resources and the minimum generation of waste are the best cure of environmental problems. Therefore, it is inevitable to adopt green manufacturing, as a "mass manufacturing" concept, which considers resource and environment factors in parallel and thoroughly at each stage of the entire life cycle of a product, so that the environment is protected and resource utilization optimized.

In addition, the digital transformation of society is in the ascendant, and it takes time for that to take effect. The industrial scenes are complex and diverse, which determines that enterprises cannot accomplish digital transformation overnight. Intelligent manufacturing emphasizes the elimination of waste from non-value-added activities, while delivering high-quality products at the lowest cost and higher efficiency. The competitiveness and profitability of an organization, if managed under the intelligent manufacturing mode, can effectively improve productivity.

To stay competitive, in the face of the current unprecedentedly fierce global competition, enterprises must design and provide better products and services, and improve their manufacturing business. In manufacturing, green manufacturing updates production processes, establishes environmentally friendly businesses, uses less natural resources, recycles and reuses materials, and reduces emissions during the production, which undoubtedly broadens the boundaries of intelligent manufacturing.

The fact is that the transformation of manufacturing modes and greater energy efficiency have always been inseparable. A century ago, Ford tested the first assembly line to assemble a flywheel magneto. He maximized human mass production capacity, and machine capacity stood out. But little known is that energy transformation is another key factor in his assembly line.

The development of electric motors has rid machines of the limitation of central power. The previous central power (like a steam engine) had to go through re-distribution of power after being driven by a gear chain, which badly restricted the layout of the machines. The distributed power from the electric motors at last allows the machines to be arranged with the highest efficiency.

The energy revolution that green manufacturing leads to is a key factor that promotes the further development of the Industrial Revolution in addition to intelligent manufacturing. When green manufacturing is combined with various digital technologies as a manufacturing mode that minimizes waste and pollution, it will not only reduce waste and pollution, but also greatly increase the productivity and profitability of the organization. Obviously, as environmental problems deteriorate and the society becomes digitalized, green manufacturing has become inevitable.

4.2.2   Green Technologies

From the concept of "big manufacturing," the entire manufacturing process generally includes: product design, process planning, material selection, manufacturing, packaging and transportation, use, scrap, and disposal. When green factors are considered at every step, green manufacturing technologies come into being.

Green Design

Traditional product design usually considers the basic attributes of the product, such as function, quality, life, and cost. It is therefore people-centered. It targets at the customer demand and starts with problem-solving, so that in traditional product design, the consumption of resources and energy and the impact on the ecological environment during the production and use of the product are often ignored.

Green design is a brand-new design concept, also known as ecological design (ED), design for environment (DFE) and life cycle design (LCD). Focused on the entire life cycle of a product, it considers the product's environmental attributes (low energy consumption, demountability, long life, recyclability, maintainability, reusability, etc.).

The basic idea of green design is to incorporate environmental factors and pollution prevention measures into product design at the design stage, to take environmental performance as the design goal and starting point, and to minimize the impact on the environment. From this point of view, green design examines the entire life cycle of a product from the angle of sustainable development, emphasizes systematic analysis and evaluation from the angle of the entire life cycle in the product development stage, and eliminates potential negative environmental impacts. Green design is accomplished through several approaches such as life cycle design, concurrent design, and modular design.

Green Materials

The selection of green manufacturing materials requires designers to change the traditional way of material selection. Under the premise of satisfying the basic functions, the materials to make the product must be green, meaning that during the product's entire life cycle, the materials help reduce energy consumption and minimize the impact on the environment.

First, the types of materials used are reduced. Using fewer types of materials not only simplifies the product structure, facilitates the production and management of components, and the identification and classification of materials, but also allows more materials to be recycled under the same product quantity.

Next, recyclable or renewable materials are chosen. The use of recyclable materials reduces not only the consumption of resources, but also the environmental pollution of raw materials during refining and processing. The body of the BMW Z1 car is entirely made of plastic and can be demounted from the metal chassis within 20 minutes. Its doors, bumpers, and front, rear, and side control panels are all made of recyclable thermoplastics produced by General Motors.

The third is that materials that can naturally degrade are used. Fuzhou Institute of Plastics Science and Technology and Fujian Institute of Testing Technology have succeeded in developing a new type of plastic film from light-controllable plastic compound additives. The film degrades into fragments within a certain period of time after use, and dissolves in the soil as food for microorganisms, thereby purifying the environment.

The last is the use of non-toxic materials. In the automotive and electronics industries, the most commonly used are solders containing lead and tin. However, lead is extremely toxic, so in recent years, it has been restricted or banned in paint, gasoline and many other products.

Clean Manufacturing

Compared with the real clean manufacturing technology, the clean manufacturing in green manufacturing is more based on the green manufacturing process technology, green manufacturing process installation and equipment.

For example, in substantive machining, green manufacturing process can be implemented in processes such as casting, forging and stamping, welding, heat treatment, and surface protection. It improves these processes, increases the product qualification rate; adopts a reasonable process to simplify the product processing flow, the processing procedures, to minimize the waste in the production, and to avoid unsafe factors. Regarding less pollutant emissions in the production, for example, to reduce the use of cutting fluid, dry cutting technology is applied.

Green Packaging

It is common knowledge that there are five elements in the marketing of modern commodities, namely product, price, channel, promotion and packaging. In a world that cares about environmental protection, green packaging is playing a more important role in sales. Green packaging refers to packaging that uses packaging materials and products that are non-polluting to the environment and the human body, but recyclable and reusable or renewable.

Green packaging means the necessity to abandon the novel consumption concepts and optimize product packaging schemes to minimize resource consumption and waste generation. First, product packaging must be as simplified as possible to avoid excessive packaging; then, it must be reusable or recyclable easily without causing secondary pollution. For example, Motorola's standard packaging boxes shrinks their size, increases their utilization rate, and replaces raw wood pulp with recycled pulp for inner packaging, thereby improving economic efficiency and enabling packaging to be "3R1D" (Reduce, Reuse, Recycle, and Degradable).

Green Recycling and Disposal

Product recycling is a systematic project, and must be fully considered since product design. The design for disassembly should be adopted to deal with the transportation issues in the recycling process, the possible changes in the state of the recycled products, and the damage or corrosion of some components. Scrapped products should be recycled and disposed of in time. On the one hand, after disassembly, components can be reused, saving lots of raw materials. On the other hand, there is less environmental pollution. Therefore, when a product is designed, it should be made easy to disassemble, and different materials can be easily separated to facilitate recycling, regeneration or degradation.

Green manufacturing itself is a kind of corporate behavior, but to some extent it has the characteristics of public goods. Therefore, green manufacturing must be turned into conscious corporate behavior. State governments need to take the first step, such as improving laws and regulations, taxation policies, and capital markets to support the effective work of environmental protection departments. However, present laws and regulations for this purpose fail to give strong support for green manufacturing activities.

Certainly, with the prevalence of carbon neutrality, the development of the global economy will sooner or later be decoupled from the higher and higher carbon emissions, and green intelligent manufacturing will become an important link to enhance the competitiveness of digital enterprises in the future. From technology to system, from predicament to breakthrough, all will be the starting point for the boom of green manufacturing.

4.3 Service-oriented Extension: Value-added Manufacturing Value Chain

At present, the value chain of manufacturing keeps extending and expanding. Manufacturing and services are gradually merging. Manufacturers are more inclined to provide customers with products, services, and their application solutions. Service-oriented manufacturing is a manufacturing mode that aims to increase the value of the manufacturing value chain. Through the integration of products and services, the full participation of customers, and the provision of production-oriented services or service-oriented production, it integrates dispersed manufacturing resources, enables the efficient synergy of their respective core competitiveness, and eventually achieves efficient innovation.

Service-oriented manufacturing is an important mode for the future transformation of the manufacturing industry. As a sustainable business mode, it will also generate huge profits to the enterprises.

4.3.1 The Evolution and Maturity of Service-oriented Manufacturing

Judging from the evolutionary path, service-oriented manufacturing, as the integration of service and manufacturing, is actually the result of the mutual movement of the manufacturing industry and the service industry towards each other.

There are two relationships between manufacturing and service: one is the customer/supplier relationship—the development of manufacturing opens up markets for many production-oriented service, such as finance, insurance, technical consulting, and logistics, and in turn these services strongly support and accelerate the development of the manufacturing industry; the other is the interdependence relationship—service depends on the product, and product sales depend on the service. For example, when a manufacturer sells a certain product, it also creates a demand for related services. When a service provider sells engineering or management consulting services, it guides customers' demand for equipment and other related auxiliary equipment.

Therefore, as the social division of labor deepens and the production-oriented service evolve, the manufacturing and service industries are integrated and interdependent. Their boundary is becoming increasingly blurred. Their integration not only leads to the change in the connotation of traditional products, but also in the traditional manufacturing organizations.

Specifically, at the beginning, enterprises used to adopt the manufacturing mode that sold a single product. They take resources from nature, process them through the manufacturing system to make products, and at last sell the products in the market through distributors or agents. The product ownership is transferred to consumers in the transaction, and ultimately obtains market value, such as the production of beverages, food, and clothing.

Next, enterprises began to sell products and render services that achieve the product functions because the products sold to customers were more complex. Therefore, to ensure the normal performance of the product, it is necessary to render matching auxiliary services, such as delivery services, maintenance and repair services for washing machines and air conditioners. As a result, services based on product functions are bundled with the products. Product functions either realize or carry the services, becoming the service carrier. The purpose of selling products is to enable customers to enjoy better services, like getting a free phone if the customer pays the customized phone fee.

Even later, services based on product functions began to have additional functions. Based on these additional functions, services beyond the common use of the product can be rendered, thus creating greater value. For example, a Texas tractor manufacturer installs probes on the tractor tracks to collect local soil information during the plowing, and transmits it to fertilizer manufacturers, so that they can produce fertilizer suitable for the local soil.

At last, the service-oriented manufacturing mode has matured. Through the integration of products and services, the full participation of customers, and the provision of production-oriented services or service-oriented production, it integrates dispersed manufacturing resources, enables the efficient synergy of their respective core competitiveness, and eventually achieves efficient innovation. It is a service based on manufacturing, and also manufacturing for service.

4.3.2 Service-oriented Manufacturing Updates the Value Chain Creation

As a new type of manufacturing mode, service-oriented manufacturing has integrated the service industry and manufacturing industry, and is updating the creation mode of the value chain, exhibiting advantages over traditional manufacturing modes.

Primarily, service-oriented manufacturing has upgraded limited service to full-cycle service. The core problem of an enterprise's traditional logistics supply chain management system is how to supply raw materials, semi-finished products and part of finished products to the demand side quickly, accurately, efficiently and at low cost. However, changes in consumption concepts and production modes have changed the relationship between consumers and suppliers. The supply chain has evolved from push-type to pull-type. Companies have changed from rendering consumers with a one-time purchase service to a full product lifecycle service.

The formation of this new type of competition and cooperation has prompted manufacturers to further outsource their businesses, thereby promoting the integration of manufacturing industry and service industry. Service-oriented manufacturing processes products in the circulation, such as the circulation processing of steel, fruits and vegetables. This better meets the customer needs, promotes product circulation efficiency, increases the profitability of manufacturers and service providers, and changes the profitability time from one-time to the entire product life cycle.

Secondly, service-oriented manufacturing creates a new mode of integration of manufacturing and service, which enables customers to transform from simple demanders to participating manufacturers. This new mode integrates resources that belong to different regions, entities, and states, and realizes the centralized deployment, use, and efficiency of the resources. Meanwhile, service-oriented manufacturing pays full attention to the needs and status of customers, introduces customers into the supply chain management system, and makes them "cooperative producers." Through production-oriented service and service-oriented production activities among member companies, the goal of maximizing customer value and corporate value is achieved. Simultaneously, the profit space and scope are expanded, and the single and single-link service income is transformed into multi-level and compound income.

Service-oriented manufacturing starts with customer demand management to integrate and coordinate the resources of the member companies in the supply chain. Customers participate in the integrated service processes of raw and auxiliary material procurement, product design, manufacturing, product packaging, logistics, after-sales service, and product recycling, so that the value of the entire life cycle of the product is increased, the value of each member company in the manufacturing supply chain is increased, and ultimately the high degree of collaboration that disperses their core competitiveness is realized.

Lastly, service-oriented manufacturing also promotes the transition from value satisfaction to demand satisfaction. The advent and development of service-oriented manufacturing are conducive to increasing the value of both products and enterprises, and also to increasing the value of the customers. The latter is realized through the deeper satisfaction of customer needs. Service-oriented manufacturing has diversified profit sources for enterprises, such as from product profit and from service profit. These enterprises can both provide products and services, and render services beyond the products, such as after-sales support, financial services, and logistics services.

Service is becoming a new battlefield for enterprises and a new way to increase the additional value of products. When obtaining a full range of services, customers feel more satisfied, thus there forms a more stable customer relationship. Service-oriented manufacturing integrates service network and manufacturing network. In this dynamic new network, resources are optimized through the market to achieve flexible manufacturing.

In general, under the globalization of manufacturing networks and service networks, the manufacturing industry and service industry mutually restrict and influence each other's development, and their integrated development is inevitable. Service-oriented manufacturing is a new business model resulting from the integrated development of the two industries, which is conducive to the transformation and upgrading of the manufacturing industry. Under the buyer's market conditions, building a supply chain suitable for the enterprise's development and performing supply chain management are the inevitable choices for enterprise development. The development of service-oriented manufacturing requires attention on the construction of the value chain system and operating mechanism to adapt to the internal and external environment of enterprise development and improve the enterprise's core competitiveness.

4.4 Industrial Internet: an Effective Path for Digital Transformation

The Fourth Industrial Revolution is the deep integration of cyber-physical systems. In this background, the industrial Internet, which can directly connect consumers and manufacturers, is the most important product of the fourth wave of industrialization. At present, the industrial Internet platform has set off an upsurge in the global manufacturing industry. As a carrier to promote the integration of a new generation of information technology and the manufacturing industry, it has become global consensus that the industrial Internet platform facilitates the transformation and upgrading of the manufacturing industry.

4.4.1 Industrial Internet to the Future

The Industrial Internet encompasses a wide range of subdivisions. From different angles, there are different understandings of the Industrial Internet.

From a technical angle, the Industrial Internet is a new industrial digital system that integrates a new generation of information communication technologies such as new networks, advanced computing, big data, and artificial intelligence, with manufacturing technologies. During a long evolution to take shape, it is a systematic integration of multiple information technologies.

From a macro perspective, the Industrial Internet supports the digital, networked, and intelligent transformation of the manufacturing industry through the comprehensive connection of all elements of the industrial economy, the entire industrial chain, and the entire value chain. It keeps breeding new modes, new business models, new industries, and reshaping industrial manufacturing and service system, so as to achieve high-quality development of the industrial economy.

Where there is the Industrial Internet, the means of production and the relations of production undergo revolutionary changes. Under the Industrial Internet, the data from the process links is collected, processed, and precipitated in the cyberspace, and eventually used back in the process links. Similar to the technological revolution in the age of steam and the age of electricity, data makes an important means of production for industrial enterprises, and communication technology becomes an important tool of production.

Certainly, understanding and developing the Industrial Internet are a gradual evolutionary process. Enterprise informatization is both the prerequisite for the development of the industrial Internet, and the internal demand of enterprises. It was General Electric in the U.S. that created the world's first financial system. During enterprise informatization, different systems are widely used in various departments of the enterprise, generating a large number of isolated information.

Extending the value chain to the users is an inevitable step for the development of the Industrial Internet. This can also be seen as a kind of service-oriented manufacturing industry. This process requires strong control over the enterprise's software development capabilities and operation mechanism for downstream users. For example, Bao Sight from Baosteel's information department, and Venus from FAW Group, not only serve their own companies, but also generate income in various projects.

Ultimately, the Industrial Internet will inevitably empower the general digitalization of the entire industry. From the perspective of cost requirements, low cost has always been the pursuit of industrial enterprises to profit more, except that the efficiency improvement of traditional physical equipment has reached the limit. The Industrial Internet's application of cloud computing and big data to transform existing machines and physical equipment will bring significant marginal improvements in costs.

For example, Uptake helped Palo Verde, the largest nuclear power plant in the U.S., to save US$10 million cost per month. The annual cost reduction was up to 20%. Another example is Qingdao Textile Machinery Factory, which relies on Haier's COSMO Plat to realize remote operation and maintenance of equipment through data collection and analysis. It saves RMB960,000 per year, shortens the downtime from three days each time to one day, and reduces direct losses of RMB640,000 every time.

From the perspective of the supply chain, the Industrial Internet proposes a new production mode to realize flexible manufacturing and personalized customization, which play a vital role in intelligent production. From the perspective of the space chain, limitation from space and resources make it difficult for traditional enterprises to coordinate multiple links. With the Industrial Internet, industrial enterprises will be able to share business information, realize real-time production monitoring, remote data collection and control, and respond in time to break the barrier of space and achieve interconnection.

4.4.2　The Industrial Internet Goes Deeper

Earlier, the industrial Internet attracted little attention. On the one hand, the industry itself has certain thresholds when compared to the easily understood consumer field. On the other hand, the industrial Internet without 5G network support is somewhat a house of cards. However, 2020 was the year when 5G commercial use was popularized, thus pushing the development of the Industrial Internet to a key node.

Meanwhile, IoT, cloud computing, artificial intelligence, and big data offer strong support to the application of the industrial Internet. The development of the IoT has enabled the collection of a large amount of industrial data including the status, identification and location of smart objects. The Internet enables data transmission. Cloud computing provides platform-based industrial data calculation and analysis capabilities. The penetration and integration of information technologies such as the Internet, cloud computing, IoT, and big data into the industrial field has contributed to the breakthrough and formation of the Industrial Internet.

Till now, the Industrial Internet has basically completed the development from concept popularization to widened practice.

The network is the foundation of industrial interconnection. The Industrial Internet requires that the internal supply, sales, and storage, production, middle-end and back-end management of the enterprise should unify the information flow of people, finances, and materials, and connect the currently independent industrial information system. Also, the information flow between the upstream and downstream enterprises of the external industrial chain should be interconnected and the overall coordination be accomplished. Therefore, the most basic requirement of industrial interconnection is the provision of the underlying support through the communication network, so that the real-time perception and collaborative interaction of different units, different equipment, and different systems take place in the information system network and the production system network.

The platform is the center of industrial interconnection. The massive amount of data generated by different units, equipment, and systems in the ecology is gathered on the platform based on the network. The nature is the construction of an efficient, real-time and accurate platform system oriented at the requirement of digitalization, being networked, and intelligentization from the big industry through emerging technologies such as the IoT, artificial intelligence, and big data, to perform data collection, modeling analysis, application development, resource scheduling, and monitoring management. This is the very core of industrial interconnection.

Security is the guarantee of the network and the platform. In the industrial Internet age, data is one of the core assets of enterprises, thus there is more emphasis placed on the information security of the system. The security of the enterprise intranet is divided into three aspects: application security, control security, and equipment security, whose general manifestation is the ability to safeguard equipment, network, and data.

Under this circumstance, the battle of the Industrial Internet has attracted wider attention. In addition to international giants such as GE, Siemens, PTC, and SAP, there are also Chinese participants such as Zhejiang sup OS, Tencent Cloud, COSMO Plat, and XCMG Information, that are responding actively to this trend, and more and more essential powers continue to drive more technological innovation in multiple subdivisions.

Rhino, launched by Alibaba in 2020, is a smart factory with industrial Internet characteristics. Rhino can access to Alibaba's huge consumer data to help small clothing companies predict which items will sell well, thereby simplifying their production plans.

Within the factory, the machines made by Rhino are equipped with smart cameras, and there are conveyor belts between the workstations. Each piece of fabric is marked and traceable, and the entire workflow is digitally recorded on Alibaba's cloud, so that business owners can track the progress from afar. By digitizing each process of the production line, Alibaba is laying the foundation for terminal apparel manufacturers to use a standardized universal operating system to run all machines.

Obviously, the Industrial Internet has gone deeper.

4.4.3 Competition and Cooperation, or Zero-sum?

The complexity of industrial subdivisions means that the deepening of the industrial Internet will be a melee. In the future, the scale of the enterprise included will be larger than ever in terms of number or subdivision. To go with competition and cooperation or zero-sum, and how to occupy the highest ground are the issues that manufacturers are faced with.

But whether competition and cooperation or zero-sum, in the vast industrial Internet, enterprises play an important role in their own positioning. On the one hand, although China is the only country in the world that includes all the industrial categories in the United Nations Industrial Classification, there is no single Chinese manufacturer able to cover all industrial segments and application scenarios.

On the other hand, the participants in the industrial Internet are different by nature, such as industrial enterprises, ICT enterprises, and Internet giants. Some start from process manufacturing, while others begin with discrete manufacturing. The technical architectures of the industrial Internet platforms they build differ, too. Finding the right position is the first step in the melee of the deepening industrial Internet.

At the same time, as Internet manufacturers evolve, data exchange, system integration and the creation of common systems become key factors. An industry is a collection of all units engaged in economic activities of the same nature. However, because the supply chain and value chain span different industries and enterprises, there are often problems such as weak enterprise connections and poor data sharing. As a result, the value of data cannot be brought to full play.

The Industrial Internet is an ecological community, where everything from elements to values is jointly completed by the community. To promote cross-industry interconnection, it is necessary to bridge between enterprises in the whole value chain and the whole business ecology, and have the data related to business operations in the whole value chain and the whole ecology flowing.

Finally, the Industrial Internet is an externalization based on the expansion of the value chain, and presents characteristics utterly different from traditional industries. Large industry gaps and intricate enterprise processes are huge challenges for the platform. Most importantly, the business logic of the Industrial Internet has evolved.

In the age of the Industrial Internet, the traditional manufacturing industry has to emerge as an all-round role of an operator. For example, can Foxconn, as an excellent manufacturer, truly become an operational company? Similarly, each company committed to playing a new role, such as Haier, Midea, XCMG, and Zoomlion, has to face the organizational change whether it is ready to face the challenge of a new business model.

The Industrial Internet is a technology integrator and requires massive convergence and integration from different disciplines, technologies and knowledge. At present, no platform possesses over 50% of the resources, technologies and capabilities the Industrial Internet requires.

Apparently, for the time being, platforms do not have a zero-sum relationship, but a competition and cooperation relationship, and cooperation outweighs competition. Only the full collaboration between the platforms and the construction of an ecological community of "sharing and collaboration" can lead to prosperity. A new opportunity for the deep application of the Industrial Internet has presented itself, which will redefine processing and production and innovate the future of the Internet.

4.5   Intelligent Manufacturing: From Technology to Model

Intelligent manufacturing is both a technology and a model.

Technically, intelligent manufacturing is based on a new generation of information technology, cloud computing, big data, IoT, nanotechnology, sensors, and artificial intelligence. Through perception, human-computer interaction, decision-making, execution and feedback, it intelligentizes product design, manufacturing, logistics, management, maintenance and services. It is the integrated collaboration and deep integration of information technology and manufacturing technology.

In the meantime, service-and-intelligent-science-oriented intelligent manufacturing based on technologies such as cloud computing, and IoT, is also an intelligent manufacturing model. It employs the network and cloud manufacturing service platform to organize online manufacturing resources (manufacturing cloud) on demand, and renders users with readily accessible, dynamic, and agile manufacturing lifecycle services.

4.5.1   Key Technologies

The intelligent manufacturing model makes the manufacturing processes high-tech, global, and transparent, in order to transform manufacturing to "intelligent manufacturing." This step is based on the upgrade of key technologies of intelligent manufacturing, including artificial intelligence, IoT, big data, cloud computing, cyber-physical system, and intelligent manufacturing execution system.

Artificial Intelligence

During intelligent manufacturing, manufacturing high technology based on technology and service innovation needs to be integrated into all production links to intelligentize the production processes and increase the value of the products. The manufacturing industry empowered by artificial intelligence shows great potential. The combination of artificial intelligence and related technologies can optimize the efficiency of various manufacturing processes. Through industrial IoT, all sorts of production data are collected, and processed via deep learning algorithms so that suggestions for optimization are made or independent optimization takes place.

Regarding the application scenarios of artificial intelligence in the manufacturing industry, there is mainly intelligent product R&D and design, the use of artificial intelligence in manufacturing and management to improve product quality and production efficiency, and the intelligentization of the supply chain.

In product R&D, design, and manufacturing, artificial intelligence not only uses algorithms to explore various possible design solutions according to established goals and constraints, and carries out intelligent generative product design, but also integrate and productize its own technical achievements to manufacture a new generation of smart products such as smart phones, industrial robots, service robots, auto-pilot cars and drones.

Artificial intelligence embedded in manufacturing links makes machines smarter, so that they do not only perform monotonous mechanical tasks, but also operate autonomously in more complex situations, thereby comprehensively improving productivity.

On the intelligent supply chain, demand forecasting is the key to the application of artificial intelligence in the field of supply value management. By better predicting changes in demand, companies can effectively adjust production plans to improve plant utilization. In addition, smart handling robots will autonomously optimize warehousing, greatly improve the efficiency of warehousing and picking, and reduce labor costs.

The IoT

The IoT controls the production process, monitors the production environment, tracks the production supply chain, and supervises the entire lifecycle of the product through information perception based on RFID technology and smart sensors, information transmission based on the integration of wireless sensor networks and heterogeneous networks, and information processing based on data mining and intelligent image and video analysis. Ultimately, it helps enterprises better understand and utilize local resources. Therefore, the IoT plays an irreplaceable role in the globalization of intelligent manufacturing.

Big Data

The advent of the global IoT has continuously produced massive amounts of data. Faced with these large-scale, diverse, high-speed, and low-value data, the key to higher intelligence of the manufacturing processes is the effective application of big data to process and integrate it so that there is transparency in the manufacturing processes, valuable information can be obtained, and intelligent analysis and decision-making can be performed to improve the adaptability.

Cloud Computing

In response to the advent of the global IoT and big data, cloud computing is based on resource virtualization technology and distributed parallel architecture to provide users with infrastructure, application software, and distributed platforms as services, so as to achieve distributed data storage, processing, management and mining. Through proper utilization of resources and services, it provides practical solutions to make smart manufacturing agile, collaborative, green, and service-oriented. On the premise that data privacy and security are guaranteed, cloud computing will gain wide acceptance from the enterprises.

Cyber-physical Systems

The cyber-physical system accomplishes real-time perception, dynamic control and information service of the manufacturing process through the organic integration and in-depth collaboration of computer technology, communication technology and control technology. As an intelligent and autonomous system, it not only obtains data from the manufacturing environment, processes and integrates it to extract effective information, but also acts on the manufacturing processes through industrial robots and other equipment according to control rules to integrate information technology with automatic technology. It is a key part of intelligent manufacturing.

Intelligent Manufacturing Execution System

The intelligent manufacturing execution system targets the needs of collaboration, intelligence, leanness, and transparency. It develops functional modules such as intelligent production management, intelligent quality management, and intelligent equipment management based on the existing traditional MES to make the production processes consistent throughout the production and intelligently control product quality. In addition, based on IoT and big data, it performs real-time remote monitoring of the production process, and emergency prediction, classification, and response, and integrates factory automation and informatization, which are the core links to make a factory smart.

4.5.2   Towards Intelligent Manufacturing

The intelligentization of the manufacturing industry is no social process that arises out of thin air, but one that receives support from previous technological accumulation, and takes advanced technologies such as artificial intelligence and a new generation of information communication technology as the turning point of industrial transformation. Digital manufacturing is the starting point for the development of intelligent manufacturing, and networked manufacturing is the transitional stage to intelligent manufacturing. The ultimate goal of manufacturing upgrades is the transformation from digitalization and networking to intelligentization. At present, the manufacturing industry is in this important transition from digitization and networking to intelligentization.

As the starting point of intelligent manufacturing, digital manufacturing is based on digital technologies. It is accompanied by the development of digital control technology and CNC machine tools. Taking product design and production links as the object of action, it emphasizes the digitalization of the production processes. From the perspective of design, digital manufacturing, compared with traditional manufacturing, mainly converts traditional manual design sketches into computer simulation design, thus greatly improving design efficiency. From the perspective of production, digital manufacturing has partially automated the production processes, which is manifested in the use of machines to complete some complex and precise processing procedures. And CNC machine tools are one of the representative technologies of digital manufacturing.

Networked manufacturing is the development stage of manufacturing intelligentization. It takes the networked age as the background. Accompanied by the development of information communication technology, it targets at the collaborative production and operation between enterprises, and emphasizes enterprise cooperation and information sharing. From the perspective of technical foundation, networked manufacturing receives support from Internet technology, which breaks through the constraints of geographic space on enterprise production and operation. The Internet enables enterprises to communicate and coordinate information. From the perspective of manufacturing purposes, compared with digital manufacturing, networked manufacturing shifts the focus from internal production of the enterprises to collaborative production between them, which is manifested in the resource sharing and integration through the network. Representative technologies of networked manufacturing include network-based distributed CAD system and open-structure-controlled processing center.

Intelligent manufacturing is the mature stage of manufacturing intelligentization. It is developed on the basis of the two necessary conditions: digital manufacturing and networked manufacturing. It takes the entire process and life cycle of product manufacturing as the object of action, and emphasizes the use of a new generation of information communication technology and artificial intelligence. And the manifestation is the application of artificial intelligence to the production operation system, so that it can perceive, decide, and execute by itself.

The core of intelligent manufacturing lies in the "intelligence." Compared with digital manufacturing, it not only fully uses computers to control, but also applies new manufacturing modes such as additive manufacturing to complete product design with special shapes and structures. This replaces the human brain and physical strength, and enables the manufacturing process to analyze, reason, and execute. Compared with networked manufacturing, it connects intelligent machines, and men and machines through the industrial Internet, which is the extension and expansion of the Internet to industry.

On the basis of comprehensive perception, ubiquitous connection, deep integration and efficient processing, intelligent manufacturing, with the support of calculations and algorithms, transforms human-oriented decision-making and feedback into a model of autonomous modeling, decision-making, and feedback of the machine or the system. The Internet offers greater possibilities for precise decision-making and dynamic optimization. Intelligent manufacturing transformed data into information, knowledge, and decision-making. It mines the hidden meaning of data, breaks the limitations of traditional cognition and knowledge boundaries, provides a quantifiable basis for decision-making and collaborative optimization, and maximizes the use of the hidden value of industrial data. This will strongly support the enabling role of the Industrial Internet in the future, thereby highly integrating people, machines, and networks.

PART 3
THE FUTURE

CHAPTER 5

New Roles, New Markets, New Rules

5.1 Industry Begins Again

5.1.1 From Industrialization to De-industrialization

The significance of industry to human life has never been doubted, but since the 1970s, there has been a wave of "de-industrialization" in the developed countries of the capitalist world.

The U.S. began to "de-industrialize" after World War II. As a traditional industrialized country that had completed industrialization and entered the post-industrialization stage before World War II, the U.S. stopped directly exporting electromechanical products and automobiles to Western Europe in order to bypass the tariff barriers of the European Community in the early post-war period, and made enormous direct investment in Europe for localized production.

The hollowing out of the U.S. industry after the war reflects the profound change of the post-war U.S. industrial structure from "real to virtual." In this process, the manufacturing industry has been shrinking, thus the new title "U.S. sunset industry." Judging from the output value ratio of manufacturing in the national economy, the U.S. manufacturing industry has exhibited an obvious decline after the war. Except for a few sectors such as electronic product manufacturing, traditional manufacturing industries such as machinery manufacturing and automobile manufacturing have suffered a long-term recession, while the virtual economy, which was supposed to serve the real economy, has been constantly expanding.

After the war, the traditional American agricultural sector continued to decline since industrialization, and the proportion of agricultural output value and employment in the national economy was insignificant; in addition to the smaller share of manufacturing in the secondary industry, other industrial sectors remained in a relatively small, stable ratio for a long time; instead, the virtual economy sector in the tertiary industry embraced rapid growth with the upsurge of "economy servitization."

In Japan, after over 20 years of rapid industrialization and economic growth in the postwar period, the proportion of the tertiary industry in Japan exceeded 50% in the late 1970s, therefore the country entered the post-industrial stage. Thereafter, through direct overseas investment and technology transfers, Japan continued to move its uncompetitive domestic industries and production links overseas. This gradually expanded from the manufacturing links at the beginning to other links in the enterprise value chain.

Judging from the industrial distribution within the manufacturing industry, the proportion of overseas enterprise in Japan's processing manufacturing industry doubled in the past two decades, and that in its material manufacturing industry exceeded 2/3 in 2007. During this period, the overseas production ratio of Japanese domestic-registered enterprises maintained a steady increase for a long time. Although the overseas production ratio of overseas registered enterprises fluctuated, it was obviously climbing as a whole.

In general, the number of overseas enterprises in the Japanese manufacturing industry surpassed half of the total number, and the overseas production ratio of the manufacturing industry calculated by the number of enterprises reached a new peak. In particular, the overseas transfer of many giant leading Japanese enterprises resulted in the overseas transfer of related supporting enterprises. But, this kind of industrial cluster transfer also deteriorated the environment of component supply and supporting product supply of large Japanese enterprises, thus causing the chain reaction risk of large, medium-sized and small enterprises successively moving overseas.

A great number of Japanese enterprises have transferred their production capacity overseas, which means that Japan's domestic capital is outflowing overseas. In addition, Japan has long taken exporting manufacturing products and overseas investment as its main economic advantages, and has imposed many restrictions on the entry of foreign capital. Many harsh restrictions in the industrial environment have made it difficult for Japan to become the first choice for foreign direct investment. Consequently, the ratio of the amount of foreign investment that Japan attracted over the Japanese economic scale has been among the lasts in the world for a long time, making it a typical country with a deficit in foreign direct investment.

From 1990 to 2004, Japan's average annual direct investment in foreign countries was about US$27 billion, while during this period, its average annual foreign direct investment was only about US$4 billion. This means that Japan's annual average net outflow of investment was about US$23 billion. This gap widened after 2005, in which Japan directly invested overseas more than US$70 billion on average, while its domestic absorption of foreign direct investment was only US$7.5 billion.

This stems from that de-industrialization is the result of enterprise behavior and capital flow in response to changes in the external industrial environment in the evolution of industrial structure. From the perspective of enterprise behavior, regardless of the increase in domestic labor costs, either the tight supply of natural resources, or the changes in exchange rates, may distance collective enterprise production and investment from local manufacturing and other real economic sectors. From the perspective of industrial capital flow, the constraints of the industrial environment lead to a decline in the return rate of industrial capital in the local manufacturing industry and other real economic sectors, so that there will be a tendency to flow to regions and industries with a higher capital return rate, and ultimately deindustrialization happens.

The stronger industrial environmental constraints are the original cause of the occurrence of de-industrialization. As industrialization deepens, it is common to strengthen industrial environmental constraints. From the perspective of the evolution of labor supply, the basic law of the population structure evolution determines that there is eventually an end to unlimited supply of labor during the window period of population opportunity, and the disappearance of the demographic dividend and the shortage of labor supply in the aging period will inevitably take place. This indicates that the industrial structure built on low labor costs is unsustainable. From the perspective of capital supply, the uneven distribution of capital among industries and the "Macmillan gap"-styled financial constraints are important causes for the deterioration of the industrial environment and operating difficulties of manufacturers. From the perspective of natural resource supply, the finite nature of natural resources determines that the industrial structure based on low-cost resource supply and environmental destruction is unsustainable, and the transformation of the industrial structure to get rid of this constraint will inevitably undergo a decline in high-consumption industries. From the perspective of the changes in the international industrial environment, as the comparative advantages of the international division of labor change, the advantages of low cost and huge market of developing countries will attract a considerable number of industrial transfers from developed countries, thus replacing the imports of products from developed countries to a great extent. So, the external demand of export enterprises in these countries will shrink. In addition, factors such as the intensification of international trade frictions, fluctuations in international market demand, and exchange rate appreciation are also important causes for the strengthening of international industrial environmental constraints.

The "hollowing out" in the strategic transformation of enterprises is the key to the occurrence of deindustrialization. There are two main types of enterprise strategic transformations that may lead to de-industrialization: the first is the strategic transformation of enterprise regional layout. As domestic labor costs rise and the market becomes saturated, enterprises will consider moving their manufacturing and other links to regions with more comparative advantages of lower costs and greater market share. The second is the strategic transformation of enterprise business structure. As the constraints of the traditional manufacturing industry environment get stronger, enterprises may adjust business structures to reduce the proportion of traditional manufacturing business, and invest in virtual economy businesses such as financial markets through centrifugal diversification strategies. If this "hollowing out" is a collective action of many enterprises, the direct consequence will be that the region and industry where they are located will experience a total decline in production capacity, and de-industrialization will break out.

The loss of industrial capital's "off localization" and "off manufacturing" is the direct cause of deindustrialization. As industrial capital that directly gains appreciation through production activities, the direct goal of its flow is higher returns on the basis of controlling risks, and the main body that manipulates its flow is the enterprises where it is located. As enterprises make strategic transformations in response to the constraints of the industrial environment, industrial capital flows, too. In line with the enterprise's regional layout transformation strategy and business structure centrifugal diversification strategy, the flow of industrial capital exhibits the trend of "off localization" and "off manufacturing." Under this circumstance, a large amount of industrial capital outflows from the local to overseas, from the real economy sector to the virtual economy sector. This implies that the local manufacturing industry and other real economy sectors of the enterprise have suffered capital loss and insufficient investment, thus the inevitable de-industrialization.

5.1.2 Industry Makes Prosperity, and Recession Too.

Now "de-industrialization" has done more harm although in Western countries it was once regarded a wise move, and an inevitable change when they were in the middle and late stages of industrialization, in which their technology and capital accumulation was sufficiently strong, and the consumption level of their residents was high.

As mentioned before, there are two main manifestations of deindustrialization: the first is characterized by the large inflow of industrial capital into the tertiary industry; the second by the transfer of industrial capital to other countries. Also, the second has different forms, and different effects on developed countries where de-industrialization occurs.

The first type of de-industrialization has two manifestations: one is the loss of productivity. De-industrialization leads labor to flow from higher-productivity manufacturing industry to lower-productivity service industry, which reduces social productivity. The other is less factor input. Relatively, the capital-labor ratio of the service industry is relatively lower, and so is its demand for capital and labor input. Therefore, as labor flows from manufacturing industry to the service industry, the derived demand for capital and labor will shrink, resulting in unemployment and slow economic growth.

In the U.S., with the decline in the proportion of manufacturing output value, a great number of laborers are "pushed out" from the manufacturing industry, and they cannot be absorbed by other industrial sectors within a short time, which leads to long-term employment in the U.S. Especially since the 1980s, there has been a drastic drop in the proportion of the U.S. manufacturing employment. This drop had to be related to the increase in labor productivity of the industry itself, but to a greater extent it was affected by the overall decline in the industrial sector.

Most of the population transferred from industry go into the service industry. But the service industry, which takes in a great number of the employed population, is divided into high-end service industry and low-end service industry. The former mainly covers financial, accounting, legal, medical, and education positions that require expertise. These positions generally offer higher incomes, but there are few of them.

Most jobs of the low-end service industry require little expertise and skills, but they pay less. The middle class of the society, namely, the blue collars, are gradually eliminated in the process of de-industrialization, thus accelerating the polarization between the rich and the poor in the society. Barriers are built between various social classes, intensifying class contradictions. So, as the trend of "de-industrialization" grows stronger, a great number of workers are unemployed, and the mobility of classes stagnates.

Regarding the second manifestation of deindustrialization, when industrial capital is transferred to other countries, industrial hollowing out inevitably occurs. Since the 1970s, countries such as the UK and the U.S. have transferred a great many high-end manufacturing industries to Germany, Japan, and South Korea. Since the 1990s, they have moved basic manufacturing industry to developing countries, especially China on a large scale. This has hollowed out the industries in the UK and the U.S., and the phenomenon of thorough deindustrialization occurred.

The lack of industrial support greatly increases the risks for a country. For example, in the UK, the service industry had increased to occupy 70% of its GDP by the beginning of this century, and the British economy had completed a fundamental transformation from production-based to service-based.

At this time, in line with the U.S. on the other side of the Atlantic, the British real estate industry boomed. When that entered the bubble stage, on the one hand, the UK industry used to be the main service target and customer of the British financial industry, but as it continued to shrink, the British financial industry invested more and more funds in developing the British real estate and purchasing U.S. subprime mortgage bonds, and eventually contributed to the real estate bubbles in itself and the U.S.

On the other hand, before the bubbles burst, the service industry represented by the financial industry in the UK benefited the most from the boom of the real estate bubbles, so there was a "win-win" situation of "common prosperity" between the real estate industry and the service industry. However, after the real estate bubbles in the UK and the U.S. burst, the situation took a turn for the worse. The British financial industry suffered heavy losses due to the double attack of its own bad mortgage debt and the U.S. subprime mortgage. The service industry, especially the financial industry and the real estate industry, which used to be regarded as the "employment reservoir," took a huge blow and had the worst situation of unemployment. The British industry with a smaller scale simply could not accommodate such a huge unemployed population, thus the soaring unemployment rate in the UK.

The unemployment rate in the UK soared, especially that a great many white-collar elites with high incomes in the financial industry and real estate industry went unemployed. Then, this triggered a chain impact on the UK's retail, tourism, catering and other service industries.

This process is also a process where more capital in the U.S. national economy was flowing towards the non-productive virtual economy. Before the outbreak of the U.S. financial crisis in 2008, the financial insurance industry and the real estate industry occupied more than 20% of the U.S. national economy. In particular, financial companies' profits accounted for as much as 40% of all companies'. However, during the same period, the proportion of traditional real economic sectors such as manufacturing and construction dropped from 50% in the early post-war years to below 30%. The hollowing out of industries under the imbalance between the virtual economy and the real economy was the culprit of the U.S. economic bubbles and the outbreak of the financial crisis. It is precisely the great harm from "de-industrialization" to the social economy of Western countries that has led to their "re-industrialization" to save the declining industrialization.

5.1.3 From De-industrialization to Re-industrialization

Obviously, the manufacturing industry is the engine of economic growth. The growth of the manufacturing industry creates more economic activities within itself and other industries outside it. It has a higher multiplier effect and extensive economic connections. Manufacturing growth creates more R&D activities than other industries of the same scale. Manufacturing innovation activities are paramount to boost productivity, and productivity growth is the source of improved living standards. And when the real economy fails to support the industrial foundation necessary for the sustainable development and prosperity of the tertiary industry, de-industrialization has to be corrected and re-industrialization be implemented.

The global financial crisis arising from the subprime mortgage crisis in the U.S. teaches a profound lesson about the re-industrialization of developed countries. In this background, Western developed countries that were once "deindustrialized" such as the U.S., the UK, and the EU countries commenced to re-examine the relationship between the real economy and the virtual economy. Manufacturing is once again valued.

In fact, "re-industrialization" is no brand new concept. It was first proposed by Amitai Etzioni, a senior adviser to the White House. The Webster Dictionary (1968 edition) defines "re-industrialization" as: "a policy of stimulating economic growth especially through government aid to revitalize and modernize aging industries and encourage growth of new ones."

The connotation of this "re-industrialization" policy no longer stays in the old scope of revitalizing and "returning to" the manufacturing industry. Essentially, it aims to develop high-end and advanced manufacturing industries propelled by high and new technologies, to upgrade the manufacturing industry, to locate growth points in the modernized, upgraded and cleaned manufacturing industry, and to lay the foundation for long-term economic prosperity and sustainable development in the future. In order to ensure the smooth implementation of the "re-industrialization" strategy, Western countries have formulated policies or introduced supportive measures to solve the predicaments in the "re-industrialization" process in a multi-pronged manner. They strive to revive the manufacturing industry through government intervention.

First, after the financial crisis, Western countries have successively formulated strategic plans to lead the development of the manufacturing industry, elevating it to a key national goal.

Barack Obama signed the "A Framework for Revitalizing American Manufacturing" in December 2009, treating manufacturing as the core of the American economy.

Thereafter, the U.S. government has set out to formulate the "Manufacturing Plan for 2040," with a focus to responding to long-term challenges from emerging powers. In August 2010, Barack Obama signed the "United States Manufacturing Enhancement Act of 2010." In previous "State of the Union Addresses," Barack Obama has repeatedly used manufacturing as an entry point to revitalize the American economy.

The British government issued the strategy report *Manufacturing: New Challenges, New Opportunities in 2008*, and announced a new manufacturing development strategy in 2009, proposing strategic ideas such as occupying the global high-end industrial value chain and seizing opportunities for low-carbon economic development. In 2010, it issued the "Forward Growth" strategy, which outlined the future development direction of industries and enterprises that play a role as the engine of economic recovery, and pointed out once again that a vibrant manufacturing industry is important to the UK.

In 2010, the Europe 2020 Strategy proposed to restore industry's rightful position, so that the industry and service industries together form the backbone of the European economic development. In the same year, the European Commission also set forth a new strategy for industrial development as an important part of the Europe 2020 Strategy to consolidate and develop the industrial competitiveness of Europe. After taking office, the President of France Nicolas Sarkozy proposed the new French industrial policies, placing industry at the core of the country's development, and set up growth targets and specific measures for the French manufacturing output.

Second, judging from the development of "re-industrialization" in various countries, green and low-carbon policies have become the main directions to revitalize the manufacturing industry.

The U.S. continued to offer stronger support for emerging industries, striving to make breakthroughs in new energy, basic science, energy conservation, environmental protection, and "smart earth." When Barack Obama took office, he implemented the new green policies, announced a new comprehensive energy plan, signed the "Green Energy and Security Guarantee Act," promulgated the "American Clean Energy and Security Act of 2009," and took a series of measures, including tax deductions, government funding, and the establishment of R&D and manufacturing centers to support the R&D and promotion of clean technologies and the popularization of clean energy equipment and products. The "American Recovery and Reinvestment Act of 2009" introduced an economic stimulus package totaling US$787 billion, in which industries such as renewable energy, energy-saving projects, and smart grids received considerable investment. In 2009, the "American Innovation Strategy" once again proposed that the government should promote the application of clean energy technologies and make enormous investments in smart grids and renewable technologies; and support the advancement of advanced vehicle technology to ensure that the U.S. stay in the leading position in this field.

The UK regards a low-carbon economy as the pillar industry of the fourth technological revolution and future development. "The UK Low Carbon Transition" plans to grant GB£ 4 million to help the manufacturing industry accomplish low-carbon transformation. The "British Low-Carbon Industry Strategy" promises to give full support to the manufacturing industry in terms of preferential policy, product procurement, educational training, standardization and capital investment. In the strategic report, *Manufacturing: New Challenges, New Opportunities*, the UK government proposed to set forth a comprehensive low-carbon industry strategy, gathering the strength of various government departments to help the manufacturing industry adapt to the low-carbon economy, with a focus on nuclear energy supply chains, renewable energy equipment, and low-carbon vehicles. On the basis of commissioning many research institutions to conduct in-depth research on the industrialization of low-carbon economy, the UK government formulated the *Low-Carbon Industry Strategic: A Vision*, proposing to take measures to build low-carbon infrastructure in the future, and make the UK a world leader in low-carbon automobile development and production. The UK government has made a series of moves to promote the development of new energy automobiles, including setting up low-emission automobile offices for better coordination and simplifying various departmental policies; granting GB£ 10 million in funding to support the development of advanced and efficient electrical systems;

launching the "United City" plan to assist major cities in setting up their charging station networks; and announcing the "Rechargeable Automobile Consumption Encouragement Scheme" to subsidize individual or group consumers who purchase new energy automobiles that meet the requirements.

The "EU Economic Recovery Plan" proposed to implement a series of plans such as "Green Partnership," "Energy Efficiency Building Partnership," "Future Factory Partnership," and "European Green Automobiles," and announced that it would invest €105 billion in developing the green economy. The EU was also focused on promoting "low-carbon" manufacturing products and processes through emission right trading, energy taxes, and green government procurements. In 2009, it issued the *European Roadmap—Electrification of Road Transport (version 3.5)*, which provided comprehensive guidance for the development of electric vehicles in the EU. In the High-Tech Strategy 2020, Germany introduced 11 "future plans" including the development of electric vehicles. In order to promote the low-carbon transformation of the automotive industry, the German government has invested enormous capital to promote the R&D of electric vehicles and the construction of related infrastructure. In addition, it launched the "National Development Plan for Electric Vehicles" and "Development Plan for Hybrid Electric Vehicle," established "National Platform for Electric Vehicles," and implemented the "Energy Innovation and New Energy Technology Research Projects," "Automotive and Transportation Technology Research Projects," and "National Hydrogen Cell Technology Innovation Project" to boost the development of these industries. To achieve the goals set in the new industrial policy, the French government focused on supporting new technologies and new energy. It formulated the €35 billion "large national debt" plan in the 2010 budget bill, granting €6.5 billion to support industries and small and medium-sized enterprises (SME), including the provision of €500 million of "green" loans to improve the production of energy-saving and emission-reduction transformation of enterprises, and enhance their competitiveness.

Finally, in the global efforts to revitalize the industry, technological innovation is regarded as the biggest driving force for the sustainable development of the manufacturing industry in the future. Countries have increased their investment to promote its "smart" growth.

The U.S. increased a US$13.3 billion investment in science and technology in the draft of the American Recovery and Reinvestment Act of 2009. The American Innovation Strategy planned to increase investment to restore America's international leading position in basic research. President Obama also promised to double the funding for basic research in the next decade. At present, the ongoing high-end manufacturing plans in the U.S. include more active research in the fields of nanotechnology, high-end batteries, energy materials, biological manufacturing, new-generation microelectronics R&D, and high-end robotics. They aim to promote the cluster development of high-end American talents, high-end elements and high-end innovation, and maintain America's the leading role in high-end R&D, technology and manufacturing. The American government has set up manufacturing policy offices as well to coordinate the formulation and implementation of manufacturing policies for the government, the industry, the academia, and the research departments, and to promote the revitalization of the manufacturing industry from various aspects such as talent training, technological innovation, tax rewards and punishments, and trade promotion.

In addition, it has also increased investment in scientific research infrastructure: the U.S. National Science Foundation has received US$200 million from the federal government to restore and strengthen the strength of the U.S. scientific research infrastructure; the U.S. Department of Energy has allocated US$327 million in the American Recovery and Reinvestment Act for scientific research, infrastructure and the update of large laboratory equipment; the U.S. National Institute of Standards and Technology has received US$360 million for scientific research infrastructure construction in the economic stimulus plan.

In the strategic report *Manufacturing: New Challenges, New Opportunities*, the UK government proposed that in order to support technological progress in manufacturing, the government's funding for scientific research increased to nearly GB£ four billion in 2010/2011, the highest ever. The British Technology Strategy Board invested GB£ 24 million more in research on high-end manufacturing. The white paper *Innovative Nation* released in March 2008 pointed out that the UK government would continue to support the 10-year science and innovation investment framework plan and increase the funding of the Technology Strategy Board. In May 2008, *Connect and Catalyze: A Strategy for Business Innovation 2008–2011* announced that the Trustee Savings Bank (TSB), together with related departments, would invest a total of GB£ 1 billion in the next three years and attract private investment of the same amount. At the end of June 2009, the British government invested GB£150 million to set up the British Innovation Investment Foundation, which drives private capital, so that GB£ one billion of venture capital was offered to start-ups and high-tech companies in the growth stage.

At the EU's International Conference of Research Infrastructure held at the end of 2008, the EU added 10 new large scientific research infrastructures at one time, increasing the number of construction projects in its scientific research infrastructure roadmap plan to 44 items in seven categories. The total construction funding reached €16.951 billion, with an annual operating cost of €2.21 billion. The German government launched the Zentrales Innovationsprogramm Mittelstand (ZIM) in July 2008 to finance science research and innovation projects. In 2009, the German federal government invested €900 million in ZIM. The federal government and the Bundesland governments funded €174 million in special scientific research infrastructure projects in German universities and colleges. The German Science Foundation granted €85 million of federal funding to large research instruments. In Germany's "High-Tech Strategy 2020," the federal and Bundesland governments have agreed that by 2015, the investment in education and scientific research will have increased to 10% of the German GDP. France's budget for large scientific research facilities in 2009 increased by €319 million, an increase of 17% compared to the previous year. The French government has also set up a special fund through the French Agency for Innovation (OSEO) to focus on supporting the science research and innovation activities of SMEs, and through the French Strategic Investment (FSI), financed the R&D and innovation activities of large- and medium-sized enterprises.

5.2   The Pan-industrial Revolution Goes Global

The Pan-industrial Revolution is a global phenomenon, and major industrial countries in the world have formulated matching strategic measures in recent years.

5.2.1   Germany: Industry 4.0

Germany is at the forefront. Industry 4.0 was first proposed at the 2011 Hannover Messe. It aims to improve the level of German manufacturing by applying new technologies such as the IoT. Germany's proposal and implementation of the "Industry 4.0" strategy are a national movement to respond to the latest technological advancement, global industrial transfer, and changes in its own labor structure.

In 2013, the German Federal Ministry of Education and Research and the Federal Ministry of Economics and Technology included the "Industry 4.0" project as one of the ten future projects in *High-tech Strategy 2020* released by the German government in July 2010. They planned to invest €200 million to support the R&D and innovation of a new generation of revolutionary technologies in the industrial field, to maintain Germany's international competitiveness, and to secure the future of German manufacturing.

Industry 4.0, which was initiated by the Merkel government and promoted worldwide, hopes to restore Germany's global leading position in the industrial field. It is also an active strategy to solve the problem of an aging population. At these heights, the fundamental goal of Industry 4.0 was to promote the upgrading of German industrial manufacturing from automated to intelligent and networked by building an intelligent production network. Focused on transforming the original advanced industry model of intelligentization and virtualization with the help of the information industry, it values smart factories and smart production, and prioritizes the formulation and promotion of new industry standards during the development, that is, the industrial integration of German Industry 4.0.

The German Industry 4.0 strategy includes the construction of smart factories based on the digital-physical system, and the procurement, marketing, and R&D of manufacturing units and enterprises in various value chain links. At the same time, it connects different enterprises to create a larger and higher intelligent production network. Through the breakthrough and application of modern manufacturing technology, it will further exert Germany's advantages in product innovation, information technology, high-end equipment and complex process management, thereby making the German manufacturing industry stronger, maintaining and consolidating its leading position in global manufacturing.

Industry 4.0 describes the basic mode change from centralized control to decentralized enhanced control. The goal is to establish a highly flexible production mode for personalized and digital products and services. This is the Fourth Industrial Revolution brought about by the IoT and manufacturing services after the first three Industrial Revolutions of machinery, electricity, and information technology. In this mode, there are no traditional industry boundaries, and various new fields of activity and forms of cooperation emerge. The process to create new value is changing, and the division of labor in the industrial chain being reorganized.

From the perspective of consumption, Industry 4.0 is a big network that covers production materials, smart factories, logistics and distribution, and consumers. Consumers only need to place an order on their mobile phone, and the network will automatically send the order and personalized requirements to the smart factory. The factory, then, purchases raw materials, designs and produces them, and directly delivers them to consumers via network distribution. To summarize this concept, it is the "interconnected factory," which connects the factory with things and services inside and outside itself through communication networks such as the Internet. It creates higher value than ever, builds new business models, and addresses many social problems.

In essence, Industry 4.0 is the integration of IT and industrial technologies. Smart factories and smart production meet the users' individual customization needs. Even one-use products can be manufactured in a profitable way. In Industry 4.0, for suppliers, dynamic business models and timely business processes make production and delivery more flexible, and enable them to respond flexibly to production interruptions and failures. Industrial manufacturing is therefore able to provide end-to-end transparency in the manufacturing process to facilitate the decision-making of selections. Industry 4.0 will explore new ways to create value and develop new business models.

In the age of Industry 4.0, network technology, computer technology, information technology, and software will be deeply intertwined with automation technology to explore new value models. The Industry 4.0 intelligent assistance system frees workers from monotonous and stylized work, so that they can focus on innovation and value-added services. Flexible work organization allows workers to better integrate their own tasks, so that their private lives and future career development will become more efficient.

By implementing the Industry 4.0 strategy, German industrial enterprises not only meet the highly personal needs of consumers, but also make timely responses and adjustments to changes in market demands and in raw material supply. At present, giants such as Siemens, SAP, and Bosch are competing in network platform technology. The German government and industry associations have established steering committees and working groups to advance the Industry 4.0 strategy. And a series of measures have been taken as well in terms of standards, business models, R&D, and talent, such as integrating relevant international standards to unify services and business models; establishing new business models suitable for the IoT environment, so that the entire

ICT industry can work more closely with machine and equipment manufacturers and mechatronics system suppliers; supporting enterprises, universities, and research institutions so that they jointly conduct research on self-regulatory production systems; and offering better training of skilled talent to better fit Industry 4.0.

5.2.2 The United States: Industrial Internet

America, as an epitome of the Third Industrial Revolution, is far ahead of other countries in the development of information technology. However, although it still remains a global leader in advanced manufacturing such as aerospace and chip manufacturing, the hollowing out of its manufacturing industry and the loss of global market share have been difficult to reverse through simple policy adjustments or business approach adoptions. Meanwhile, the country is also challenged with a similar demographic threat. International consumers' requirements for product customization and diversification have also, both internally and externally, prompted it to use its advantages in the information industry to transform its manufacturing industry.

The U.S. started the National Strategic Plan for Advanced Manufacturing in 2012, thoroughly interpreting the idea of "re-industrialization." It proposed to develop advanced digital manufacturing technologies, including advanced production technology platforms, advanced manufacturing processes, and design and data infrastructure, in order to encourage innovation, reshape the industrial structure through IT, and activate traditional industries. This "top-down" reshaping of the manufacturing industry from the CPU, systems, software, Internet and other information terminals via big data analysis differs from Germany's "bottom up" idea of starting from the manufacturing industry and using IT to transform it.

In the face of the Fourth Industrial Revolution, the U.S. has adopted a representative measure of the Industrial Internet. In 2012, the American Industrial Internet strategy was officially promoted to a national strategy. The Industrial Internet involves a vast physical world based on machines, equipment, clusters, and networks. Able to be combined with connectivity, big data, and digital analysis, it aims to revolutionize a series of key industrial fields, thereby promoting their successful transformation and upgrading.

The core content of the Industrial Internet strategy is mainly reflected in the information supply network, and its technical model mainly involves Internet technology, big data, cloud computing, and broadband networks. By implanting dissimilar sensors in different links of the manufacturing, it performs continuous real-time sensing and data collection. With the help of data, it accurately and effectively controls the industrial links, thus achieving the goal of efficiency improvement.

The Industrial Internet in the U.S. is focused on the integration of elements, especially that of the results of the Internet. There are three major elements: industrial intelligent machines, advanced analysis, and staff; the element of intelligent machines is to effectively connect machines, equipment, and networks with the help of sensors, controllers, and software applications, and to continue to promote the efficient integration of "information," an important production element; the advanced analysis element performs an overall grasp of the operation mode of machines and huge systems, thereby prompting data to get 100% ready for technical integration; the staff element mainly builds a real-time connection of staff in different workplaces to effectively realize more intelligent design, operation, maintenance, high-quality service, and a safety guarantee.

From the perspective of the developmental trend of the Industrial Internet, first the manufacturing industry becomes service-oriented, which promotes the transformation of manufacturing from pure product manufacturing to service-oriented manufacturing; then, there is more customization, that is, the expansion from mass-produced products to customized products; the third is the decentralization of the organization. Affected by the continuous integration of the Internet, the industry has gradually exhibited the characteristics of a decentralized organization; the last is the cloudification of manufacturing resources. Digital companies build industrial clouds and gradually integrate design, supply, procurement, and manufacturing on a certain platform.

At present, the Internet and information communication technology giants in the U.S. have joined hands with traditional manufacturing leaders. GE, AT&T, IBM, Intel, and Cisco Systems, and have jointly established the Industrial Internet Alliance. Over 100 companies and institutions from the U.S., Japan, and Germany are its members. They have together decided on the basic framework of IoT standardization, analysis, applications, and innovative practices.

For example, in order to promote the Industrial Internet, GE established the General Electric Software Center in 2011 as a base for software development and promotion. It jointly funded the establishment of Pivotal, a software company, with EMC to handle the big data management business in the Industrial Internet. In 2014, Pivotal began to provide external customers with the Industrial Internet core data analysis software Predix.

5.2.3 China: Made in China 2025

At present, as a manufacturing giant, China has not yet become a world-class manufacturing power. It is in the parallel development stage of "Industry 2.0" and "Industry 3.0." China's manufacturing development lacks both Germany's strong foundation in traditional industrial fields and America's advanced technologies that lead the global development of information technology. Also, it is necessary for China's manufacturing industry to address the basic issues such as product quality improvement, stronger basic industrial capabilities, and manufacturing upgrading and transformation.

In response to the Fourth Industrial Revolution, on March 5, 2015, Premier Li Keqiang pointed out in the government work report that "Made in China 2025" should be implemented to accelerate the transformation from a manufacturing giant to a manufacturing power. This enlivened the domestic industry, and attracted wide attention from the international economic community. On May 8, the State Council printed and distributed the "Made in China 2025" document, comprehensively promoting and implementing the strategy of building China into a manufacturing power.

"Made in China 2025" adheres to a path of new industrialization with Chinese characteristics. It aims to promote the innovation and development of the manufacturing industry, improve quality and efficiency, accelerate the in-depth integration of a new generation of information technology and manufacturing, and further intelligent manufacturing, so as to supply major technical equipment for economic and social development and national defense construction, strengthen basic industrial capabilities, improve comprehensive integration, enhance multi-level and multi-type talent training system, promote industrial transformation and upgrading, cultivate a manufacturing culture with Chinese characteristics, and realize the historical leap from a manufacturing giant to a manufacturing power.

As a country with an enormous population, to become a manufacturing power, it is necessary to for China to realize the following goals: the first is to expand the industrial scale so that there is a mature and complete modern industrial system and the Chinese industrial scale holds a considerable proportion in global manufacturing; the second is to optimize the industrial structure until the manufacturing industry has a more perfect industrial structure, high level of basic industries and equipment manufacturing, high proportion of strategic emerging industries, and many powerful multinational enterprises and a legion of vibrant and innovative SMEs; the third is to ensure good quality and efficiency, which are manifested in advanced production technologies, excellent product quality, high labor productivity, and occupying high-end links in the value chain; the fourth is the sustainable development capabilities, including the strong independent innovation capabilities and gradual growth in scientific and technological leadership capabilities, and the ability to achieve green and sustainable development as well as a satisfying level of informatization.

In order to attain the goal smoothly, "Made in China 2025" mobilizes all social forces to formulate strategic measures and action plans with "innovation-driven, quality first, green development, structural optimization, and talent-oriented" as the basic principles.

From the perspective of innovation driven, China is actively constructing intelligent manufacturing engineering and the manufacturing innovation system. Focused on intelligent manufacturing, it is accelerating the integration and development of a new generation of information technology and manufacturing technology, developing smart products and equipment, promoting the digitization, networking and intelligentization of the production processes, cultivating new production modes and industrial models, and comprehensively improving the level of intelligence in enterprise R&D, production, management and services.

Meanwhile, it is necessary to establish a technological innovation system that takes enterprises as the main body and that integrates production, education, and research. Revolving around the major common needs of the transformation and upgrading of key industries and the innovation and development of key areas, a number of manufacturing innovation centers will be built to carry out the R&D of industry foundations and common key technologies, the industrialization of results, and talent training.

From the perspective of quality first, Germany solved the quality problem as early as in Industry 2.0. China must be determined to solve the quality problem if it wants to become a manufacturing power. It should emphasize the promotion of the new technological revolution in improving quality, and better address quality problems innovatively with informatization of Industry 4.0, and intelligent new technologies and new methods.

For example, in terms of a strong industrial foundation, weak industrial foundation capabilities including basic components, basic processes, basic materials, and basic industrial technologies foundations (collectively referred to as the "4B") are restricting the quality improvement and innovative development of the Chinese manufacturing industry. To lay a solid industrial foundation, it is necessary to coordinate the "4B" development, strengthen the cultivation of the "4B" innovation capabilities, and promote the coordinated development of whole-machine enterprises and "4B" enterprises.

Also, to substantially improve the quality of industrial products, China is required to comprehensively raise stronger quality awareness, improve quality control technology and quality management mechanism, strengthen and popularize advanced manufacturing standards, further brand building and produce brand products with independent IP rights, and continue to enhance the brand value and the image Chinese brands.

From the perspective of green development, China insists on taking green development as an important focus of becoming a manufacturing power. It walks the development path of ecological civilization, transforming itself from extensive manufacturing with high resource consumption and high pollutant emissions to resource-saving and environmentally-friendly green manufacturing. It increases the efforts in the R&D and promotion of advanced energy-saving and environmental protection technologies, processes and equipment, and in the acceleration of the green transformation and upgrading of the manufacturing industry; it actively promotes low-carbonization, recycling, and intensiveness, and improves the efficiency of manufacturing resources; strengthens the green management of the product life cycle and strives to build an efficient, clean, low-carbon, and recyclable green manufacturing system.

From the perspective of structural optimization, China insists on setting it as the main direction of becoming a manufacturing power. It vigorously develops strategic emerging industries, promotes traditional industries to march towards the mid-to-high end, and propels the transition from production-oriented manufacturing to service-oriented manufacturing. The spatial layout of industries is optimized, the construction of modern enterprises is strengthened, and a number of industrial clusters and enterprise groups with core competitiveness are cultivated.

The key to pushing the manufacturing industry from giant to powerful lies in high-end equipment. It is necessary to concentrate superior forces to promote equipment innovation in superior areas and strategically contested areas, in order to achieve major breakthroughs in the new generation of information technology industry, high-end CNC (computer numerical controlled) machine tools and robots, aerospace equipment, marine engineering equipment and high-tech ships, advanced rail transit equipment, energy saving and new energy vehicles, power equipment, agricultural machinery equipment, new materials, biomedicine and high-performance medical equipment.

In terms of being people-oriented, China takes talent as the foundation of becoming a manufacturing power, walking a talent-oriented development path. It strengthens the overall planning and classification guidance for the development of manufacturing talent, establishes and improves scientific and reasonable selection, employment, and education mechanisms, reforms and perfects the school education system, builds and consolidates the adult education system, and accelerates the training of technical, operations, and management, and skilled talent urgently needed for the development of the manufacturing industry, so as to build a large manufacturing talent team with a reasonable structure and high quality.

5.3 "Lighthouses": The Modern Factories

Lighthouses are part of navigation. On the sea, they guide the ships, illuminate the darkness, and keep the ships safe. In the pan-industry age, there are a group of factories that illuminate the path and set examples for the pan-industry. They are called "lighthouses."

"Lighthouses" are a new role in the pan-industrial age, and a demonstrator of "digital manufacturing" and "globalization 4.0." They verify the hypothesis that all-round improvements in production value drivers can lead to new economic value—whether it is resource productivity and efficiency, flexibility and responsiveness, product launch speed, or the ability to customize the product to satisfy the customers, all can generate new economic value.

"Lighthouses" highlight the globalization of pan-industrial manufacturing in the future. A German company may set up a factory in China while a Chinese company may do that in the U.S. Innovation does not distinguish between regions and backgrounds. From the procurement of basic materials to the processing industry, and to high-end manufacturers who cater to special needs, the industry varies greatly and encompasses everything. It also means that companies of all sizes have the potential to innovate and stand out in the pan-industrial wave, whether it is a global blue-chip company or a local one with less than 100 employees.

"Lighthouses" value collaboration and open doors to thousands of visitors every year because they know that the benefits of a collaborative culture far outweigh the threats from competitors. They can inspire other enterprises, help formulate strategies, improve worker skills, collaborate with other enterprises participating in the revolution, and manage changes throughout the entire value chain.

"Lighthouses" are regarded as "lighthouses" because they are at the forefront of wide application of new technologies, and are the benchmark and pioneers of intelligent manufacturing; because they deeply integrate digitalization and manufacturing, and in terms of business processes and management systems, they have specific applications and substantial innovations; and more because the common laws that they explore to intelligentize manufacturing can provide valuable enlightenment and a reference for other manufacturers.

5.3.1 The Birth

"Lighthouses" are factories that have successfully pushed the manufacturing technologies of the pan-industrial age from the pilot stage to the large-scale integration stage, and achieved financial and operational benefits.

In 2018, the World Economic Forum and McKinsey & Company conducted a comprehensive screening of over 1,000 leading manufacturers from all walks of life around the world. After on-site visits and records, the Fourth Industrial Revolution Expert Committee conducted a thorough review and certified 16 "lighthouses," including the factories of Bayer, BMW, Bosch, Danfoss, Fast Radius of which UPS owned shares, Foxconn, Haier, Johnson & Johnson, Phoenix Contact, P&G, Rold, Sandvik Coromant, Saudi Aramco, Schneider Electric, Siemens, and Tata Steel. In 2019, 28 new members were added to the global lighthouse network. In 2020, the World Economic Forum announced 10 more lighthouse members in the network, bringing the total number of members to 54.

"Lighthouses" are the most advanced production sites in the world today. They have overcome the typical challenges faced without exception by enterprises, such as excessive proof of concept, slow promotion, lack of integrated business cases related to technologies, over deployment of isolated solutions, and creation of countless isolated data islands, etc. Ultimately, they generate a revolutionary effect, resulting in agile and continuous improvement.

"Lighthouses" are "lighthouses" owing to their production system innovation and end-to-end value chain innovation.

Production system innovation refers to that an enterprise expands its own competitive advantages through excellent operations. It aims to optimize production systems and improve operational efficiency and quality indicators. Under normal circumstances, enterprises will first pilot in one or several factories before wide promotion.

Obviously, by enhancing productivity and flexibility for transformation, the manufacturing industry can achieve inclusive economic growth and global benefits. However, for most organizations, to do that, they will face severe challenges. McKinsey's research findings show that in the popularization of the Fourth Industrial Revolution technologies in production, more than 70% of industrial enterprises are still stuck in the "pilot dilemma." Only 29% of them have adopted active measures to apply these technologies on a large scale, and more (30%) have not yet launched trials or are about to.

Most industrial enterprises are stuck in the "pilot dilemma." Not only are they clueless about a comprehensive transformation, but the experiment progress is also slow. If the enterprise cannot move from pilot to scale, the transformation will be meaningless. There are many causes for this situation, such as: lack of a long-term digital strategy, insufficient internal construction capabilities of the organization, limited scale expansion, lack of attention and support from senior management, lack of leadership from business departments, lack of empowerment from ecological partners, etc. It all boils down to the fact that the enterprises do not regard them as their strategic transformation.

The Pan-industrial Revolution is characterized by interconnection and transparency, intelligent optimization, and flexible automation. When the pan-industry is discussed, the discussion is not just about a technical concept, nor the transformation at the factory level, but an upgrade of the entire organizational structure, a strategic layout in the new industrial age.

To stand out in the pan-industrial era and successfully transform is not as simple as applying one or a few technologies, but requires overall consideration. First, the transformation must be driven by business departments; second, it is necessary to break traditional concepts and integrate multi-functional working methods; third, processes are reengineered and employees continue to be trained; fourth, a digital execution engine is built to promote the transformation of the organization from top to bottom; fifth, a scalable industrial IoT infrastructure is set up; and finally, enterprises in various fields are united to establish an ecosystem of technical partners.

An end-to-end value chain innovation creates new business for enterprises by changing the economics of operation. Innovation is deployed throughout the value chain to provide customers with new or improved value propositions by launching new products and new services, and high customization, smaller batches or shorter production cycles. Enterprises will first implement innovation and transformation in a certain value chain before extending their experience and capabilities to other departments.

On the end-to-end value chain, there are as many as 92 best digital use cases for the global lighthouse network, including supply network docking, end-to-end product development, end-to-end planning, end-to-end delivery, customer docking, and sustainability. In addition, compared with an enterprise that adopts traditional operating systems, the new operating system created by "lighthouses" has a higher return on investment, which is a significant competitive advantage.

5.3.2   To Build "Lighthouses"

Economies of scale are the key to building a "lighthouse." The daily metabolic rate of a 30 g mouse is one watt while that of a 3 kg cat is only 32 watts. When the size of organisms becomes larger, the efficiency of energy utilization increases, and there is less energy each cell needs to metabolize per second. Therefore, the metabolic rate does not multiply by 100, but only 32. This increase in efficiency from the increase in scale is the economy of scale.

"Lighthouses" can take the lead under the pan-industry because they have learned and successfully broken through the "pilot dilemma" and achieved economies of scale. Their economies of scale began with the change in operating modes. Throughout the transformation, they simultaneously made efforts in four aspects: business processes, management systems, personnel systems and industrial IoT, and data systems. They conducted in-depth innovations in the operating system, and gradually increased digital tools to make $1 + 1 > 2$.

The key to successful large-scale deployment lies in the systematic application of the five driving factors, namely an agile manufacturing mode, an agile digital studio, a technology ecosystem, the industrial IoT, and a scalable data infrastructure.

Agile Manufacturing Mode

Agile manufacturing adopted highly advanced information technology to quickly allocate various resources available, respond to the ever-changing business environment, and adjust product structure in a timely manner to meet the diversified user needs to the greatest extent. It creatively forms a dynamic alliance of enterprises—virtual enterprises, so that they can flexibly and quickly respond to market changes, and there is sufficient flexibility in the entire manufacturing production system in terms of technology, management, personnel, and especially the organization.

In order to achieve agile manufacturing, in addition to information integration and process integration, there must also be enterprise integration, which is to select partners for a specific product to form a dynamic enterprise alliance. It aims to make full use of the design resources, manufacturing resources, and human resources of the alliance, to complete the information integration and process integration within the alliance, and to quickly launch new products to the market.

Based on the principles of agility, enterprises can innovate and transform in an iterative manner in order to achieve scale development. Agile modes enable organizations to continuously collaborate, change management models, predict technical limitations, and break technical bottlenecks. For the "lighthouses," this means rapid iteration, quick failure, and continuous learning. They have to create a minimum viable product (MVP) within two weeks, and bundle use cases for multiple rounds of rapid transformation (each round lasts for several months).

Agile Digital Studio

The establishment of an agile digital studio creates a space for the development team to manage and operate based on agile working methods. This atmosphere broadens the participation of employees and supports innovation at all levels within the enterprise. The co-location of translators, data engineers, ERP system engineers, industrial IoT architects, and data scientists is a requirement to maintain agility, as is the guidance from product managers and agile mentors. This combination delivers results fast and accomplishes rapid iteration.

For example, Unilever Dubai Personal Care Site (DPC) has clarified its own competitive advantages by improving costs and customer responsiveness. It was committed to building an agile digital studio in order to create value within a short time, which helped it discover its growth potential, and improve its cost and customer responsiveness. Meanwhile, it also promoted the reorganization of work processes so that they focus on empowerment, sustainability, and value creation directly related to performance.

Factory management realized that deploying numerous third-party solutions would bring many challenges and put pressure on the enterprise costs. Therefore, they immediately formed an internal team, in which a former process engineer was appointed as the digital project leader. The team consisted of a group of engineers and technicians who would participate in these projects while continuing to fulfill their original duties. The numerous applications the team developed and delivered have guided the daily arrangements of the factory operators. Therefore, the entire team strictly followed the following principles: all applications are based on a shared data pool; an open-source platform design is used; an audio-visual user interface is provided; and mobile technology is adopted as much as possible for development.

DPC has also established cooperative relationships with some start-up companies. As the logistics force of DPC, these start-ups flexibly adjusted solutions according to the needs of DPC. For example, the DPC operation and maintenance team cooperated with a start-up to create an easy-to-use but powerful cloud operation and maintenance management system. The deployment cost of this software is not high, and it costs extremely little to order it. Fundamentally, the driving factors in the digitalization of DPC did not require substantial capital investment, such as connecting appropriate resources, confronting daily challenges, and deploying and maintaining a great number of internal solutions. On the contrary, the real input came from employees. Enthusiastic and united, they worked as a team to come up with innovative solutions.

Technology Ecosystem

The technology ecosystem consists of a series of technology-supported relationships. In other words, new types of collaboration, including data sharing, are built on digital infrastructure. Leading organizations are increasing the number of partners to make themselves more capable. These relationships are unique because enterprises can exchange massive amounts of data and collaborate on technology platforms to promote exchange and consumption. Compared with the traditional concept of technical solutions and data as competitive advantages, this is a remarkable change.

"Lighthouses," well aware of the benefits of network effects, have launched such cooperation with suppliers and partners from all walks of life to build a new type of open collaboration relationship including data sharing, thereby creating a technology ecosystem.

The Industrial IoT

The development and adoption of IoT is an important link of the implementation of intelligent manufacturing. Although manufacturers have automated sensors and computers for decades, these sensors, programmable logic controllers, and hierarchical structure controllers are separated from the upper management system. They are an organizational mode based on a hierarchical structure, so the system lacks flexibility. Because it is designed for specific functions, the functions of various industrial control software are relatively independent and the equipment adopts different communication standards and protocols, so that there is automation isolation between various subsystems.

The Industrial IoT adopts a more open architecture to support a wider range of data sharing, considers global optimization from the perspective of the whole system, and supports the perception, interconnection, and intelligentization of the entire manufacturing life cycle. In terms of system architecture, the Industrial IoT adopts a scalable, service-oriented distributed architecture. Manufacturing resources and related functional modules are virtualized and abstracted into services, providing business process applications for the entire manufacturing life cycle through an enterprise service bus. The various subsystems of the Industrial IoT have the characteristics of loose coupling, modularity, interoperability and autonomy. They can dynamically perceive physical environment information and take intelligent actions and reactions to promptly respond to user needs.

Texmark Chemicals, a client of Hewlett Packard (HP), operates a petroleum refinery in Galena Park, Texas, USA. It is one of the world's largest manufacturers of DCPD (dicyclopentadiene). In addition, it is a fee-based manufacturer that produces specialty chemicals for contracted customers.

Texmark is an important link in the supply chain of chemical products. As dangerous goods subject to strict supervision are often used, safety is Texmark's top priority. For Texmark, the Industrial IoT is the key to employee safety, production and asset management systems. This requires it to combine sensors with advanced analysis software to generate insights, automate the environment, and reduce the risk of human error.

The IoT can benefit Texmark's production flow in many ways. Obviously, professional manufacturing requires more than just a set of "universal" solutions. The IoT requires strong connectivity to collect data through various IoT devices. But this kind of networking performance must be cost-effective, but the cost of connecting the entire factory to the network through wired means is extremely high. In addition, all technologies installed in the Texmark factory must be rigorously controlled and meet the company's safe operation standards; the equipment running around Texmark must never be a source of fire. Another big challenge for Texmark is data delay. It takes time to transmit data, and the transmission time on the IoT is usually calculated in seconds. Therefore, Texmark needs a set of IoT architecture that does not need to transmit device data.

In order to meet these challenges and realize the benefits of IoT architecture, Texmark decided to adopt a multiple-phase development and deploy a set of end-to-end IoT solutions.

The first and second phases have laid the foundation for digitalization by realizing "edge-to-core" networking. The cost of deploying a wireless solution is approximately 50% of that of a wired network. For edge analysis, Texmark has deployed a set of industrialized solutions to provide enterprise-level IT capabilities at the edge. Also, it has upgraded its factory control room to achieve a seamless "edge-to-core" connection, and integrated its operations technology and IT into the same system. The third phase continues to develop on the basis of these technical solutions to provide support for Texmark's use cases, including: predictive analysis, advanced video analysis, safety and security, interconnected employees, and full life cycle asset management.

According to factory manager Linda Salinas, the IoT architecture of Texmark not only can sort out data, but also reveal the interconnection status of the entire factory. Like a living factory, it knows how to operate, and automatically flag when it encounters problems, so that it can intervene in a timely manner.

Scalable Data Infrastructure

To build a "lighthouse" requires a redesign of the current IT system and update the new generation of technical functions so that the selected industrial IoT architecture is sufficiently adaptable and can withstand future tests. Although early use cases are still applicable to a traditional IT infrastructure, most old facilities fail to meet the requirement for latency, data flow, and security of advanced use cases. For example, many traditional enterprises admit that they are not ready for more advanced use cases, and it doesn't hurt to postpone the modernization of IT and data architecture.

The factories certified as "lighthouses" take different approaches. They are well aware of the importance of speed, and deeply understand that it is crucial to break the technological barriers from time-consuming projects and provide employees with an infrastructure that enables them to complete innovation within a few weeks. Therefore, this architecture has been deployed in the early stages of digital transformation (or even before that) in order to achieve exponential expansion throughout the organization.

Fast Radius is an American additive manufacturer that combines a powerful digital back-office and digital planning to create a scalable data infrastructure and make information transparent between functional departments to address inefficiency.

This analysis platform can collect data information of the entire manufacturing processes, and employ a variety of machine learning algorithms to provide specific feedback for all links of the value chain, so that the culprit that causes problems in different functional departments can be found and taken down. This flexible platform is accomplished through an open communication protocol between all factory sensors and a central cloud data storage.

This kind of data feedback loop promotes the improvement of the design plan, thereby reducing quality problems and the times of rework. In addition, the application of a digital twin has made remote production come true, extending the coverage to all factories. This helps to assign specific tasks to specific factories while logistics and production capacity are optimized. Since its implementation, the inventory of Fast Radius has dropped by 36% and 90% less time for the products to reach the market.

5.4   Insight into "Lighthouses"

5.4.1   Procter & Gamble's Rakona Plant: Cost-leading Growth

The Rakona plant of Procter & Gamble (P&G) is a "lighthouse" with outstanding achievements. It is an epitome of large multinational companies. There are pan-industrial manufacturing technology clusters and manufacturing mode clusters deployed on its plant level and group level.

The Rakona plant stands 60 kilometers away from Prague. Built in 1875, it has a long history. During the period of communist rule, it was once a state-owned asset before P&G acquired it in 1991. Every day, it produced about four million bottles of dishwashing liquid, dishwashing powder, and fabric enhancers. As the public demand for cleaning products shifted from powder to liquid, from 2010 to 2013, P&G's sales fell sharply. Despite economic pressures and various uncertainties, the Rakona plant still hoped for a resilient and sustainable future.

Faced with this challenge, it launched a new project in order to significantly reduce costs and attract new business. The project succeeded in continuously lowering the cost of the Rakona plant and in stimulating demand. At last, it decided to expand between 2014 and 2016. To make this expansion work required digitization and automation, and through pan-industrial empowerment, it could predict and solve emerging needs.

Dual-factor Drive

Rakona plant director Aly Wahdan, once stated: "We need urgency to come up with attractive solutions. We will actively explore this vision in the plant, include all employees in this journey of innovation, and enhance competitiveness by minimizing losses." Based on this, the Rakona plant successfully conducted pan-industrial innovations with the support of the dual core drives of using an external digital environment and improving the skills of employees.

Regarding the use of the external digital environment, Rakona's leadership discovered that the internal team lacked the necessary skills to promote the innovation of the Fourth Industrial Revolution, so it took corresponding measures. They acquired external knowledge of digitalization and automation in a variety of ways, including establishing direct connections with Prague universities, collaborating with start-ups, and allowing students with an education in digitalization to work with Rakona employees through exchange programs.

As for improving the skills of employees, the Rakona plant has launched a project open to all employees, aiming to deepen their understanding of new technologies such as data analysis, intelligent robots, and additive manufacturing, and to bring them closer. In this way, employees have acquired some professional skills, and new positions such as "cyber security director" have been set up. This "pull," different from the top-down "push," was the key to creating an inclusive innovation culture. It aimed to make the entire organization 100% involved in the digital transformation.

Top Five Use Cases

In fact, all "lighthouses" choose different use cases, but they receive similar benefits. For P&G's Rakona plant, the top five use cases were digital orientation setting, process quality control, a general packaging system, end-to-end supply chain synchronization, and modeling and simulation.

The digital direction setting is a set of digital performance management systems that can have an impact on both technologies and management systems. It not only solved the difficult and time-consuming problem of data collection, but also avoided decision-making based on inaccurate data points. The digital direction setting tools directly display real-time KPI on the touch screen of the production workshop, thus allowing users to investigate data at multiple levels to understand the factors behind that determine performance and find out the root cause of the deviation.

In addition, this system can also be used to dispatch and track frontline employees. As a result, the operation of the entire system becomes more stringent, and the process reliability and overall equipment effectiveness (OEE) is improved as well. Having adopted high-frequency testing and iterative agile development methods, the entire plant can successfully execute digital transformation.

Process quality control solved the problems that existed in the previous manual sampling process, because it could not guarantee that the quality of every product in the same batch meets the standard. If a deviation is noted later, the entire batch was scrapped and reworked. Also, process quality control addresses product launch delays related to laboratory analysis.

Currently, Rakona's quality control is based on a real-time analysis of all sorts of data from multiple sensors, which monitor pH, color, viscosity, and activity level. If a deviation is noted, the corresponding production line will be shut down so that the front-line staff can inspect the batch quality and prepare a report. This system, developed by P&G, is the first of this kind in the industry. With the promotion of IT/OT integration, P&G first tested it on the new production line and then extended it to the entire system. With repetitive manual labor reduced, employees are more relaxed.

As a result, the rate of rework and complaints have been halved, and there has been a significant drop in scrap and quality inspections. As zero-time product launch has come true, the output time has been shortened by 24 hours. At present, the use case is deployed on all production lines.

For the general packaging system, Rakona has developed a unified packaging system called UPack, which makes it easy to implement any formulation changes even when the production line is in operation. In the past, only when the production line was shut down completely could the changes be executed. This means that front-line staff had to spend a lot of time manually setting up the machines and waiting.

This system that P&G developed on its group level has now been deployed to all packaging lines. It fully integrates sensors, cameras, scanners, and packaging materials, thus able to inspect and verify the status of each area. In contrast to the paper data recording mode, UPack performs automated production line inspection, so that each area of the packaging line can stay in different stages (such as start-up, production, no-load, or conversion). Based on the recipe data stored in the system and process quality inspection, UPack also automatically configures the machine. It greatly reduces the handover tasks of front-line staff, shortens the handover time by 50%, and shrinks minimum order quantity by 40%.

End-to-end supply chain synchronization includes the scrapping of excess products after each event, inventory capital constraints, slow launch to market, and difficult and time-consuming manual supply chain analysis. Based on the ever-changing user needs, P&G has continued to improve its products, thus the final invention of this globalization tool—end-to-end supply chain synchronization. It is applied to the management level of the plant. Every department uses it to coordinate with the central planning team. P&G uses this Internet-based tool for analysis, modeling, and simulation to clearly observe the end-to-end situation of the supply chain. By simulating the situation of the entire supply chain under different conditions, the problems are identified and the agility of the supply chain is improved. The tool can display all the information of the supply chain at each node, and performs in-depth analysis and optimization of each product and production line. It can also serve as a benchmark between P&G's different plants and production lines for comparison. After applying this tool to all products and production lines, inventory has gone down by 35% in three years, and inventory efficiency up by 7% from the previous year. It has also reduced the number of returns and out-of-stocks, and improved the speed to go on market when new products were launched.

Modeling and simulation can understand the possible impact of production line adjustments and reduce the test cost of production settings, so that new product defects can be identified before operation and the high cost of error correction avoided. This use case involves the systematic application of descriptive and diagnostic modeling and some predictive pilot modeling. These modeling applications are all aimed at cultivating normative modeling capabilities. Sample modeling application includes manufacturing output related to new product launches (like recommending SKU (stock keeping unit) allocation and the number of storage tanks to the production line), selecting the best conveyor speed, determining the ideal packaging size, simulating changes in the production line before actual implementation, and predicting in advance failure and identifying the root causes. Intuitive models and engineer's operability are important driving factors. This approach can avoid failure in the first place, thereby improving product design, perfecting problem statements, and optimizing test methods.

Under these innovations, Rakona improved productivity by 160% within three years; customer satisfaction by 116%; customer complaints were down by 63%; overall plant costs down by 20%; inventory down by 43%; unqualified products down by 42%; and conversions down by 36%. At present, Rakona is still marching towards a more ambitious goal.

5.4.2 Rold: Small Scale but Big Future

Elettrotecnica Rold Srl of Italy represents the SMEs of the "lighthouses." It has successfully deployed different pan-industrial manufacturing technologies and manufacturing modes in one factory. With only 250 full-time employees, Rold is an SME that specializes in the production of washing machine door locks. The company's plant in Cerro Maggiore is a small organization, but after the systematic application of digital manufacturing technologies, its productivity and quality have experienced great improvements, which means that even if the scale of investment is limited, it is possible to perform pan-industrial innovation through cooperation with technology providers and colleges and universities with the help of ready-made technologies. For example, Rold only hired three programmers.

Before the digital transformation, Rold was under tremendous pressure because its own production capacity failed to meet the growing demands of international customers. In addition, there were other problems in the factory, including poor judgement of its actual performance and recording data on paper in a decentralized manner. Front-line staff had to spend a lot of time writing reports, and most business decisions were made through assumptions and experiences, which seriously harmed operational efficiency. In this context, Rold initiated a digital transformation.

Rold achieved improvement by changing management and communication methods. Drawing lessons from a series of projects aimed at changing organizational thinking and improving skills, the company has invested in manpower to encourage employees to embark on a digital journey along with technologies. In the process, Rold must cultivate a sense of inclusiveness rather than exclusiveness among them so that they can realize that the use of digital technologies in the production workshop can create huge opportunities.

Rold encourages suppliers, customers, top management, and front-line managers to participate in industry-related activities. It has conducted some mentoring interactions with designers, engineers, employees, and external researchers with a focus on issues such as problem solving, creativity, management transformation, communication and innovation. At the same time, the company has also built relationships with industry and innovation partners and contacted representatives of international universities and associations. For example, it cooperates with middle schools and colleges to establish a technical internship model, partners with international and domestic universities, and sends employees to attend international training and related conferences.

In terms of organization and governance, Rold is committed to equipping all types of employees with the necessary skills to innovate, including training software developers and electrical engineers to model, develop, and implement IoT applications, and teaching industrial engineers digital integration skills. These measures can complement the digital transformation projects approved by the board of directors, and enable employees at all levels of the organization to embrace the arrival of the pan-industrial age.

Regarding the application of manufacturing technologies, Rold first integrated machine alarms, clarified priorities, and performed data analysis to solve problems, thereby improving overall equipment effectiveness (OEE) because front-line staff can view fault information of specific machines and customize alarms on smart watches and interactive displays.

Then, the use of digital dashboards to monitor OEE helped with real-time monitoring of production resources distributed in different factories. This enabled front-line staff to find the cause of the downtime.

Next, Rold's sensor-based manufacturing KPI report digitized any type of production machines, and it could also collect production data in real time to build dynamic interactive dashboards.

Also, the company's cost modeling was used to determine whether they should do independent manufacturing or external procurement. It combined the granular data collected by the IoT equipment in the production workshop with business-intelligent tools to increase its own accuracy.

Lastly, Rold shortened the time for the new products to go on market through 3D additive manufacturing rapid design prototyping, and accomplished several innovations. This strengthened its relationship with universities and obtained funding for research projects. As a result, Rold won the "2018 Electrolux Innovation Factory Award."

A series of innovations in manufacturing technologies and manufacturing modes have helped Rold make huge financial and operational achievements. Between 2016 and 2017, Rold's total revenue went up by 7%–8%, and the driving force behind was the 11% increase in OEE.

5.4.3   Haier: Industrial Interactive Platform

The Haier Group, founded in 1984, has been growing steadily for over three decades. It has become a large international business group with a high reputation at home and abroad. Its product range has expanded from refrigerators only in 1984 to a product group with over 15,100 specifications in 96 categories including white goods (appliances such as refrigerators, ovens, freezers, and washing or drying machines), brown goods (electronic items such as televisions, DVD players, stereos), and beige goods.

As the world's fourth largest manufacturer of white goods and one of China's most valuable brands, Haier has built localized design centers, manufacturing bases, and trading companies in over 30 countries around the world, with a total of more than 50,000 employees. It has evolved into an enormous multinational business group.

In 2007, the Haier Group achieved a global turnover of RMB118 billion. Nineteen products under the Haier brand, including refrigerators, air conditioners, washing machines, televisions, water heaters, computers, mobile phones, and smart home integration, have been rated as China Top Brands. It has successfully entered the ranks of world-class brands, and its influence is rapidly rising with the expansion of the global market. In 2007, its overall market share in the Chinese home appliance market exceeded 25%, maintaining the number one position.

Driven by User Needs

In the wave of the Pan-industrial Revolution, Haier applied digital technologies to closely connect user experiences with daily operations. In 2016, it established COSMO

Plat (www.cosmoplat.com), a Chinese Industrial Internet platform with independent property rights that involves users throughout the entire process. COSMO Plat is driven by user needs. Through user participation throughout the entire process of demand interaction, product design, production and service, it has accomplished the mass customization of "integration of production and sales."

Primarily, COSMO Plat gathers the effective needs of a considerate number of users, and attracts designers, module vendors, equipment vendors, logistics providers, and other resources, thus forming strong user and resource advantages. For example, the open innovation sub-platform can realize the innovative interaction among users, expert communities, research institutions, and technology companies, and offer first-class innovative solutions; the intelligent manufacturing sub-platform can enable the order interaction among users, equipment vendors, and manufacturers, thus making the process transparent.

Secondly, as a multinational giant with over 30 years of manufacturing practice, Haier covers seven major business links, including interactive customization, open R&D, digital marketing, module procurement, smart production, smart logistics, and smart services. Therefore, COSMO Plat transforms the small data of user requirements and the big data of intelligent manufacturing into replicable mechanism models, microservices, and industrial apps to help improve the efficiency of corporate upgrades.

Finally, the COSMO Plat platform is able to help users drive intelligent manufacturing capabilities. With a complete set of upgrade capabilities including standardization, modularization, automation, informationization, and intelligentization, it enables the interconnection of people, machines, and materials, and realizes user-order-driven production whose quantity is only one; secondly, it has the ability to integrate the industrial chain. The integration of the upstream and downstream design, intelligent manufacturing, service, and other resources of the enterprises leads to the transition from customized products to customized services, such as the upgrade from recreational vehicle (RV) customization to smart travel customization.

In general, COSMO Plat enables consumers to design and order a tailor-made product. The customer performance monitor screens the data in real time to analyze product performance and reports all signs of deterioration to the manufacturer. When a customer contacts Haier to complain about a product, the data engine retrieves performance data from the customer's product serial number to determine the root cause of the problem and action is taken. This helps to track the responsibility. If the product defect comes from the workshop workers, the workshop bonus system will record it in their personal profiles. If it comes from the component itself, the performance of the component will be inspected to determine the proper solution to prevent another occurrence. COSMO Plat has made remarkable achievements: product quality has gone up by 21%, labor productivity up by 63%, delivery cycles down by 33%, and employees' ability to monitor customer performance up by 50%.

Based on the concept of open multilateral co-creation and shared ecology, Haier has assembled more than 3.9 million suppliers, connected over 26 million smart terminals, and rendered data and value-added services to 42,000 enterprises. The COSMO Plat solution has been successfully replicated in the electronics, equipment, automotive and other industries, leading the formulation of international mass customization standards to assist enterprises to transform and upgrade.

Driven by the demand to generate greater value and the mechanism to share, COSMO Plat's self-reinforcing and self-expanding ecology continues to grow, forming an ecological attraction upon resources from all parties. In order to replicate and promote mature technologies and models to a wider range, the COSMO Plat mass customization model has been applied in 11 regions and 20 countries.

Mass Customization of COSMO Plat

COSMO Plat's mass customization solution covers seven major links in the entire process.

User interaction solutions: From limited selections to unlimited co-creation, users become designers. Based on their diverse interactive community, their fragmented and personalized needs are integrated, and plans continue to interact and be iterated, so that they can review and select the plan that best meets their needs, and verify the feasibility through the technology that integrates virtuality and reality to ensure the high precision of the enterprise manufactures from the start.

Iterative R&D solutions: From closed to open, the world becomes a platform R&D department. With R&D centers around the world as contact points, 3.2 million first-class innovation resources around the world are connected, and innovative resource support is made available to enterprise transformation through services such as definition and release of requirements, search and matching modules, project docking modules, and negotiation support modules.

Precision marketing solutions: From finding customers for products to finding products for users, precise docking is promoted. Based on SCRM member management and user community resources, demand digitization, business digitization, and data parallelization are conducted to model and analyze data, so as to sketch user portraits and perform label management, to achieve precision marketing, and to render enterprises service from product demand forecasts to user scenario forecasts.

Modular procurement solutions: From component procurement to modular procurement, suppliers participate in front-end design. Component vendors become module vendors. There is a change in the provision from components according to drawings to modular solutions to interactive users; the company is transformed from closed component procurement to an open platform for modular vendors to interact with each other, and from internal evaluation to user evaluation; the relationship between the two parties is changed from rivals into win-win competitors, from buyer and seller into both stakeholders, thus helping the suppliers to realize on-demand design and modular supply.

Intelligent manufacturing solutions: From mass production to mass customization, users participate in the manufacturing process. Through the COSMO Plat-IM module, the user orders go straight to the factory. Through mobile terminals and PC terminals, it is feasible to run an online office for the whole manufacturing process, to make the data of the quality process transparent and traceable, to allow users to deeply participate in the manufacturing process, and to improve the precision, quality, and efficiency of the manufacturing process.

Smart logistics solutions: From the factories to users' homes, real orders, direct delivery, and delivery on demand take place. Intelligent multi-level cloud warehouse solutions, main line assembly solutions, regional visualization distribution solutions and the final 1 km delivery solutions are made available to visualize the whole process of logistics from order placement to order completion. And user evaluation is used to drive the whole process of self-optimization.

Smart service solutions: From maintenance services to smart services, services create value. First, when the product becomes an intelligent network device, it can continue to render users with ecological value-added services. For example, refrigerators can perform pesticide residue detection on food and push healthy recipes. Second, with the support of cloud data, it is feasible to render the remote maintenance service of self-diagnosis, self-feedback, and self-reporting, and support enterprises to drive continuous product iteration through user data.

On top of that, COSMO Plat has set up 11 interconnected factories around the world, achieving a 71% non-storage rate. By exporting socialization capabilities overseas, it has empowered 15 types of industry ecology, including clothing, food, housing, transportation, and health care, thereby making the lives of the global users better. COSMO Plat leads in formulating the international standard for mass customization modes, which is the first time for a Chinese enterprise to dominate an international standard of the manufacturing mode. The World Economic Forum certified the interconnected factories empowered by COSMO Plat as the world's first "lighthouses," among which China has only one. This sets a new benchmark for the transformation and upgrading of the global manufacturing industry.

5.4.4 "Rhino Factory": A New Digital Manufacturing Mode

On September 16, 2020, Rhino Intelligent Manufacturing, the world's first new manufacturing platform created by Alibaba, made its debut. It is a digital intelligent manufacturing platform that had been confidentiality operating for three years. Oriented at SMEs, it first explores the apparel industry with piloted cooperation with over 200 small- and medium-sized business owners on Taobao. On the same day, Alibaba's new manufacturing "Project No.1"—Rhino Intelligent Manufacturing plant started operations in Hangzhou, Zhejiang Province in China. Moreover, it made the cut to be one of the ten new lighthouses that the World Economic Forum added to the "Global Lighthouse Network." So far, Alibaba has become the only transboundary tech company on the list of 54 lighthouses in the global lighthouse network.

The Five Cores of Rhino Manufacturing

First, the brain is necessary. In traditional manufacturing cooperation, branding businesses generally use their own experience and pre-judgment of the market to decide on various indicators of products. Taking the apparel industry as an example, it is necessary to predict the style, material, popular color, output, etc. With decisions already made on various indicators, branding businesses convey the design and requirements of new products to the factory, which produces the products as required. When the production is completed, the finished goods are delivered to the branding businesses for sale.

This is the traditional mode where sales are based on production. Per the scheduled output, a sales plan is made. Under this mode, branding businesses are exposed to relatively higher risks and greater uncertainties. Once there is a forecast error, there will be two scenarios: one is the overestimation of the market, which leads to poor sales, large inventory backlog, high costs, and tight cash flow; the other is the underestimation of the market. Products are sold out too quick for the supply chain capacity to keep up; traditional factories cannot respond to order demand in a timely manner, resulting in a shortage of goods, damaged brand image. But underestimation is still better than overestimation of the market situation.

Rhino Intelligent Manufacturing relies on Alibaba's massive shopping big data to run big data analysis and forecasts, so as to provide cooperative businesses with future product sales trends. It replaces the subjective forecasts of branding businesses with data forecasts to improve forecast accuracy. By predicting the business categories, branding businesses send demand orders to Rhino Intelligent Manufacturing. This completes the transformation from "production-based sales" to "demand-based sales," which significantly reduces the pressure on their inventory and cash flow.

Second, the digital process map. Generally, in the traditional clothing manufacturing communication mode, after the brand designer finishes the design drawings, the factory produces a sample, improves it, and determines the final form. Next, it starts small-scale trial production, inspects the production line's operation capacity of the product, and confirms the yield before mass production.

Rhino Intelligent Manufacturing claims to adopt a 3D simulation design. Through digital simulation, it delivers the design to the maximum extent and reduces the offline manual communication costs. In fact, this type of 3D simulation is common in the equipment manufacturing industry, and software such as ANSYS and pro-e can perform multiple functions such as motion simulation and performance simulation. However, this requires relatively higher assembly tolerances and better dynamics. Additionally, simulation is more suitable for tests with high costs and long cycles. In the apparel industry, it remains debatable whether the cost of simulation software is lower than the test costs and whether the best effect is achieved.

Third, the intelligent dispatch center. The optimization of the intelligent dispatch center is mainly manifested in the production end of the assembly line. The assembly line of a traditional garment factory is linear, and the hanging of clothes is a one-way circulation. Because of the different efficiency of factory workers, it is easy for congestion to happen in an assembly line. For example, the link of printing clothes operates fast, but the following link of button sewing is slower, which leads to plenty of clothes to accumulate in the button sewing link, thus the congestion. Also, every worker has a different efficiency on sewing buttons.

Officially, Rhino Intelligent Manufacturing claims to adopt a global plan to coordinate and intelligently optimize and match production capacity. The actual plan turns out to be abandoning the previous linear one-way assembly line operation, but adopting a spider-web-style hanging equipment. Through artificial intelligence and IoT behind, production capacity is automatically allocated to vacant stations, thus greatly improving production efficiency. In terms of workers' salary, "work more, paid more" is implemented to encourage them to work more efficiently.

Fourth, the regional central warehouse supply network. After the traditional factory negotiates with the branding businesses, one of them is going to find a matching raw material supplier. However, this traditional mode has drawbacks. First, there is the time cost and channel cost of finding suppliers. In the huge raw material market, it is difficult to find a cost-effective supplier. Second, traditional factories generally have long-term fixed raw material supplier partners, with whom they share a stable interest relationship. It is difficult for branding businesses to shake it. In the end, they have to compromise with the factories on product presentation and cost control. Third, the types and quality of raw materials in traditional factories are limited, thus unable to meet 100% of the needs of branding businesses. Fourth, unspoken procurement rules prevail in the industry, and it is common to supply inferior materials than agreed, swap inferior materials over fine ones during the supply, and take rebates. Branding businesses are unable to strictly control these links, thus the higher implicit costs.

The regional central warehouse supply network is the credit endorsement for raw material reservations and sales ends, and to a certain extent builds a communication platform for raw material procurement. Obviously, it is of great necessity to build a raw material communication and sales platform for the B-side. With the endorsement of the Alibaba platform, there is inevitably less price difference from middlemen. Meanwhile, data analysis can partially give purchase and material preparation reference, which is conducive to solving the problem of raw material supply.

Fifth, flexible intelligent factory. In order to increase profit margins and economies of scale, traditional factories are generally focused on conquering famous brand customers. Big orders mean stable production capacity and low moral hazard. Therefore, small and medium brands present the characteristics of small orders and high moral hazards. Once there is a cash flow crisis, it is difficult for traditional factories to get the final payment. Therefore, most traditional factories show little enthusiasm towards small- and medium-sized brands.

According to the official advertisement of Rhino Intelligent Manufacturing, Rhino production can take a minimum order of 100 pieces, and deliver the finished product in as soon as seven days. This is undoubtedly a positive thing for the small- and medium-sized brands. In traditional manufacturing factories, the minimum order is usually 5,000 or even 10,000 pieces. Different minimum order quantities correspond to different product unit prices. Small- and medium-sized brands, because of risk considerations and uncertainties in the market, dare not order in large quantities. Rhino Intelligent Manufacturing has greatly lowered the threshold for cooperation, which is conducive to the steady development of small and medium-sized brands in their early development. The first group of brands to cooperate with Rhino Intelligent Manufacturing are the 200 small- and medium-sized businesses on Taobao, and the cooperation is going to be expanded in the future.

Rhino Factories Adapt to the Apparel Industry

At the first annual meeting of Global Lighthouse Network of the World Economic Forum on the evening of September 17, 2020, Zhang Yong, Chairman and CEO of Alibaba Group, stated his opinions on new manufacturing: "It is in line with Alibaba's values; our starting point of new manufacturing is customer demand."

"Lighthouses" are regarded as the leaders of the Fourth Industrial Revolution. They are selected by the World Economic Forum and McKinsey & Company from thousands of manufacturers around the world. The World Economic Forum expects "lighthouses" to take the lead in setting new benchmarks for global manufacturing and jointly illuminate the future of global manufacturing. In the past three years, leading global manufacturers such as BMW, Schneider, Saudi Aramco, Siemens, and P&G have been certified as lighthouses. Alibaba entered the list when it was the third expansion of the global lighthouses. It holds two firsts: the first tech company from the Internet industry; the first time for the apparel industry to be there with high-tech value-added industries such as energy, electricity, semiconductor memory, and automobiles.

Before its debut, Rhino Intelligent Manufacturing had been quietly studying manufacturing for three years. In the three years of staying low-key, it had fully mastered the flexible manufacturing mode of small orders but quick return in the apparel industry, and gradually opened those capabilities to SMEs and manufacturing factories.

Obviously, the apparel industry is a typical perceptual consumer industry with seasonality. In the competitions of the apparel industry, rapid response is the key. Due to the development of the Internet and the rapid exchange of information, fashion trends change quicker than before. Small orders but quick returns have become the pursuit of garment manufacturers. Rhino factories' ability to quickly complete orders matches this characteristic of the apparel industry. During garment production, the preparation of technical materials or information determines a factory's responsiveness.

In addition, the cutting, component production, sewing, and afterfinishing of the production processes show flexibility, which by nature is how to match the relationship between order demand and production capacity. Rhino factories adopt a bridged product-o-rial system (the so-called checkerboard transport line in the media) with flexible material routing capabilities.

The World Economic Forum evaluates that Alibaba's new manufacturing platform "combines powerful digital technologies with consumer insights to create a brand new digital manufacturing mode. It supports end-to-end on-demand production based on consumer needs. By shortening delivery time by 75%, reducing inventory by 30%, and saving water consumption by 50%, it helps small businesses become more competitive in the fast-growing fashion and apparel market."

Breaking the game open with new manufacturing, Rhino Intelligent Manufacturing started with the apparel industry. The industry that once gave birth to the spinning Jenny that opened the curtain of the First Industrial Revolution falls behind in the technological applications of the Fourth Industrial Revolution. The Rhino Intelligent Manufacturing Platform relies on Alibaba's cloud computing, IoT, and artificial intelligence to take full advantage of a flexible manufacturing mode that is characterized by small orders and fast responses. More importantly, these systematic capabilities are not for their own use, but open to SMEs and manufacturing factories on the platform. This coincides with the original intention of the World Economic Forum to build more lighthouses.

In this regard, Hou Wenhao, a senior McKinsey expert and head of the Tsinghua McKinsey Digital Competence Center, believes that Rhino Intelligent Manufacturing is more than a lighthouse, but a complete ecosystem reconstructed via cloud intelligent manufacturing. Its front-end connects users and SMEs, while the back-end bridges suppliers of raw materials, IoT, and logistics, thus truly becoming a "lighthouse network" / "lighthouse ecology."

5.5 Human-centered Future Production

The process of the Pan-industrial Revolution, whether it becomes intelligent, agile, or informationized, and flexible, is no simple "technical substitution," but an organization mode that makes the assembly line work that has been extremely refined or even differentiated since the Industrial Revolution "people-oriented" again. This means that the future production in the pan-industrial age will inevitably be human-centered. It is a new challenge to human society to realize people-centered production in the pan-industrial age.

5.5.1　Improve Innovation Ability

In 2017, President Xi Jinping pointed out that "In the age of the Internet economy, data is a new production factor, a basic and strategic resource, and an important productive force." Obviously, in digital times, the three elements of data productivity, namely laborers, means of labor, and subjects of labor, will experience tremendous changes.

Among them, laborers, as the most active part of productivity, have undergone fundamental changes in the characteristics of their own production activities, their structure, and the relationship between man and nature at different stages of human society's development. In the agricultural society, human beings conducted limited development of land resources through strenuous manual labor to survive. Since the industrial society, the invention of machines liberated laborers from strenuous manual labor.

The Pan-industrial Revolution has extensively popularized smart tools, thus bringing the ability and level of mankind to transform and understand the world to a new peak. Not only have machines replaced workers to handle extensive strenuous manual labor, but data productivity has taken over lots of repetitive mental work. As a result, people can create more material wealth with fewer working hours. This also means that in the age of data productivity, innovative talents that create new products, new services or new business models will dominate the market.

When the development of machine civilization becomes the inevitable trend of modern society, the coordinated development of human civilization requires to focus on the contribution of human beings to the division of labor—supplementing the rationality of machines instead of trying to compete with them. This requires workers to cultivate innovative thinking, the awareness of challenging authorities, and even irrational ideas, not because irrationality is positive, but because it is a supplement to the rationality of machines. Only by doing so will we differ from the machines, and it is this differentiation that creates value.

The enterprise culture of "lighthouses" often attaches great importance to the participation of front-line staff and encourages them to think about how to innovate, so as to promote the successful and continuous application of technologies.

The Rakona Factory of P&G in the Czech Republic convenes regular meetings to discuss problem solutions, where the source of the problems is identified and digital solutions are formulated to avoid losses. If the source of the loss is at the top of the Pareto curve, the loss will be dealt with first with the enterprise inputting relevant digital resources. Next, the data scientists will work with the operator to find the cause, design a solution, apply the agile working methods to quickly build the minimum viable product (MVP), and evaluate the preliminary output results. In the process, the MVP is continuously tested until the confirmed elimination of the loss source.

Ford Otosan (a Turkish automotive company) built an agile talent development team that combined human resources, production processes, and vocational training to help employees develop skills related to the Fourth Industrial Revolution, such as innovation and data utilization. The team members have experienced the change from being only responsible for measuring business to analyzing automatically generated data, that is, completing administrative tasks, obtaining indicators, and submitting new ideas. They actively participated in the whole process from the evaluation and selection of new technologies to cooperation with engineers and experts, and the development of new technologies.

5.5.2   Cultivate Compound Talent

The pan-industrial age is a new age where industry is integrated with emerging technologies and crosses fields and there are various types of technologies. With the maturity of technologies such as the IoT, data collection, and cloud computing, the cost of computing resources continues to drop. Many ten-year-old problems, like the difficulty to replicate and promote, have been solved, and a large number of new technologies have begun to fuel the development of the pan-industrial industry age. Under these circumstances, it is necessary to cultivate matching pan-industrial management talent.

Deloitte, the global accounting and consulting company, proposed six drivers of manufacturing competitiveness in the Global Manufacturing Competitiveness Index, among which talent is one of the most recognized drivers. People are the theme of the Internet age while in the IoT age all things are connected. This will also stir up a greater technological wave and bring more opportunities. With the implementation of pan-industrial ideas, manufacturing practitioners will change from looking for professionals who are only good at one single discipline to compound management talents who can integrate multiple disciplines and professions.

A survey in 2019 showed that 55% of "lighthouses" were cooperating with universities or other educational institutions to absorb knowledge and talents. In addition, 71% of "lighthouses" were setting up internal colleges and competence centers for capacity building. Moreover, in addition to on-the-job training, enterprises' skill training programs also offered job rotation, temporary tasks, exchanges, and internships to help employees acquire new skills.

For example, P&G adopted the technical means to attract new talent, especially international talent, in internships and rotation programs. Petrosea, with an efficient and dynamic content delivery model, incorporated new methods such as augmented reality, virtual reality, and digital learning centers into its on-the-job training, and improved the efficiency of digital learning for employees through gamified skills training.

Skilled talent can even have a powerful impact on the overall competitiveness of a nation. America attaches great importance to cooperation among governments, enterprises, and universities. Cincinnati, where the Second Industrial Revolution started, was the cradle of American manufacturing. It accommodates the headquarters of GE Aviation and P&G. The University of Cincinnati is also known as the "West Point" of industrial big data analysis. Since 2000, it has been committed to the industrial application of industrial big data analysis and predictive maintenance technologies. Simultaneously, it established IMS (Intelligent Maintenance System) under the initiative of NSF (National Science Foundation).

IMS pays more heed to the cultivation of talents' compound skills and to the integration of industrial multi-scenario applications and theoretical innovation. It integrates industrial big data into the discipline of mechanical engineering, and divides the courses into four main parts: data technology, analysis technology, platform technology, and operation technology. IMS believes that only when cross-field technologies are integrated and there is the mutual reference between industries can more generalized technologies be developed.

Germany believes that without vocational education 4.0, there would be no German Industry 4.0. Despite the changes in the traditional apprenticeship education, it still aims to combine theories with skills, that is, enterprises and schools cooperate in education, tailoring the skill training necessary for the future according to the needs of the enterprises.

The German education system adheres to the CPS concept and applies digital technologies throughout education, because the changing manufacturing industry requires talent to transform traditional industry service providers and machine operators into multi-faceted industrial technical talent able to make man-machine dialogues throughout the production process.

In China, the 2016 New Engineering Course blew the horn for multidisciplinary talent training in Chengdu. Following that, Fudan Consensus, Tianjin University Action, and Beijing Guide consolidated the foundation for the construction of new engineering courses. Compared with the old engineering courses, the new engineering courses emphasize the practicability, intersectionality and comprehensiveness of the disciplines, especially the close integration of new technologies such as information communication, electronic control, and software design with traditional industrial technologies.

In addition, competitions such as the China Industrial Big Data Innovation Competition and the National Intelligent Manufacturing Entrepreneurship Competition held in the past couple of years have also enabled theories to be integrated with practices through the combination of competition and education. They have become an effective way to cultivate talent. In the first Chinese College Students' Mechanical Engineering Innovation and Creativity Competition of Tongji University, the participants' competition results could be directly converted as part of their GPA for the master's program.

Higher engineering education has transformed from a "technical paradigm" to a "scientific paradigm" first, and then to a practice-oriented "engineering paradigm" that always aims at the future. Compared with traditional engineering talents, the emerging industries and the new economy in the future need high-quality compound "new engineering" talents with strong engineering practice ability, strong innovation ability, and international competitiveness.

5.5.3   Adjust the Organizational Structure

Sharpened tools make the job easier. An excellent organizational structure is a powerful tool to support businesses of all sizes in the commercial society. From a holistic perspective, an effective organizational structure unites individual strengths, drives them toward the same goal, and achieves the effect of 1 + 1 > 2. From an individual perspective, a reasonable structure is also related to all aspects of daily work.

As the emerging technologies expand and deepen their impact on the industry and the society, the production mode and management of the industry have begun to change profoundly into more specialized and systematic. In this case, the society has entered a critical period of comprehensive and major changes, and the current organizational structures of various enterprises are also faced with impacts.

For example, in 2020, Coca-Cola announced organizational restructuring; Fliggy started a new round of organizational restructuring; Kuaishou issued an internal letter announcing organizational restructuring. Internet giants such as Tencent, Alibaba, and JD also performed major organizational restructuring. Xiaomi went even further, making five organizational adjustments within less than eight months since its IPO.

For enterprises, the organizational structure is so paramount that its adjustment is not only an important link connecting the next stage of development, but also illuminates the core issues that the enterprise will have to solve in the future. It determines an enterprise's success.

5.5.4   To Innovate the Industrial Organizational Structure

Every system has a structure. It consists of structural elements and their relationships. The system is built to meet the needs of stakeholders, who often have their own concerns, which a series of structural perspectives address and correspond to.

From the perspective of management, management focuses on integrating resources well, giving full play to the value of resources, and rendering value-added services. This process has to deal with interpersonal relationships, so as to maximize the value of resources. These relationships are presented through the structures and processes. The construction of an organizational system, including that of mechanisms, processes, and structures, is the foundation for the development of an enterprise.

In other words, the organizational structure, as all members of the organization, can lay the foundation for achieving organizational goals and conducting division of labor and collaboration. The structural system formed in terms of job scope, responsibilities, and rights is a management mode designed to improve efficiency.

Before building a structural system, the first job is to identify as many stakeholders as possible. In the new industrial age, with the rise of intelligent, flexible, and agile manufacturing, business parties, product managers, customers/users, development managers, engineers, project managers, testers, operation and maintenance personnel, and product operators are all possible stakeholders.

In this process, it is necessary to deeply understand the concerns of different stakeholders and provide structural solutions. Meanwhile, their concerns may conflict, such as the management (manageability) and the technical side (performance), the business side (more, faster, better, and cheaper) and the technical side (reliability and stability). At this time, a flexible structure is needed because it can balance and meet the needs of different stakeholders.

Once the stakeholders are identified, the establishment of an organizational structure begins. In the industrial field, the organizational structure has gone through a long-term evolution. Inside it, the business structure is the productivity, the application structure, the production relations, and the technical structure is the means of production. The business structure determines the application structure, which needs to be adapted to the former. And as the business structure continues to evolve, the application structure is built at last on the basis of the technical structure.

The monolithic architecture is similar to the primitive clans, where there is a simple internal division of labor and no external connection between the clans. Distributed architecture is like the feudal society, where each family is self-sufficient, but there is a small amount of exchange between families. A service-oriented architecture (SOA) resembles the industrial age. Enterprises provide various finished goods and services. All for one and one for all.

Businesses are often relatively simple at the beginning, such as purchase-sale-stock. At this stage, internal users are given a simple management information system (MIS) for data addition, deletion, modification, and inspection. A monolithic application suffices. As business deepens, each part of purchase-sale-stock has become more complicated, and in the meantime, customer relationship management has been added to better support marketing. Consequently, the depth and breadth of business have increased. At this time, it is necessary to split the system according to the business into a distributed system.

Under the ongoing Fourth Industrial Revolution, technologies represented by the IoT, cloud computing, and artificial intelligence are promoting the embedding of manufacturing systems in the governance and organizational structures of enterprises. With a novel production mode, employees have to not only execute instructions, but also make decisions on the spot. On-site workers become knowledgeable employees who can participate in product design and adjust the production processes.

An enterprise's governance structure that matches the novel production model must be able to stimulate the enthusiasm of workers with knowledge. In this case, the employee's interest demands and participation in decision-making are better reflected. At this time, the common governance structure and organizational structure that adapt to the digitalized industry will gain more weight. Therefore, innovating the organizational structure become inevitable for all enterprises.

5.5.5 The First Signs of the Organizational Structure Adjustments

In the Fourth Industrial Revolution, the "embedding" of manufacturing technologies and manufacturing system means that while the R&D of manufacturing technologies are widened to promote the breakthrough and application of modern manufacturing technologies and manufacturing system, more attention must be paid to supporting technologies, modern production management modes, knowledgeable employee training, and the improvement of enterprise organizational structures and operating mechanisms that have a strategic complementary relationship with modern manufacturing technologies and manufacturing system.

Only by strengthening the cultivation and promotion of complementary capabilities while developing modern manufacturing technologies can modern manufacturing technologies be transformed into real competitiveness of the products, the enterprises and the industry. A survey of America's flexible manufacturing system revealed that up to 20% of the equipment in the flexible manufacturing system in the early 1990s had not been put into use, mainly because both the enterprise organizational structure and employee competence failed to match with the new equipment.

It is foreseeable that the Fourth Industrial Revolution is inevitably accompanied by changes in product innovation, management, and business models. "Lighthouses," regarded as the leaders of the Fourth Industrial Revolution, were selected by the World Economic Forum and McKinsey & Company from thousands of manufacturers around the world. As the leading force the World Economic Forum expected them to be, "lighthouses" have set a new benchmark for global manufacturing. Among all "lighthouses," 71% are adjusting their organizational structure in one or more ways to vigorously promote the transformation of the Fourth Industrial Revolution.

"Lighthouses" add data-based tasks that are different from the past to traditional positions, and set up new positions at the factory or group level to meet the growing demand for data, programming, and digitization. Simultaneously, they have also changed the old organizational structure model where IT was separated from operations, established a cross-function team to focus on digital deployment, and built a model where data scientists and data engineers work closely with front-line staff.

For example, Schneider Electric in Batam, Indonesia has set up a new digital transformation department to simplify the rotation of project managers. The project manager is usually on duty for one year, but the time is flexible and adjustable. This position is designed to focus on the digital transformation projects selected in the roadmap. Also, it pays great attention to the joint development of solutions with SMEs, production line leaders, and managers in related fields.

Bosch launched projects at the corporate level to promote digital transformation, taking both methods (agile working methods, transformational leadership, etc.) and technologies (IT, data analysis, etc.) into consideration. Nokia organized a small team of 10 experts responsible for the development and application of new technologies. Similar to the engineering team, it is committed to promoting the Fourth Industrial Revolution innovation and large-scale deployment through cross-functionality.

The manufacturing revolution that occurred in the factories is part of the overall strategic change of the enterprises. Leading global manufacturers have never neglected the investment in complementary assets and capabilities while increasing that in advanced manufacturing technologies. When discussing the strategy of recovering the advantages of the manufacturing industry in the U.S., the GE president also pinpointed that in addition to developing advanced manufacturing technologies and material technologies, there should be more human capital innovation, including forming a more flexible employment system through the negotiation between capital and the labor unions, and training highly skilled workers with modern knowledge.

The innovation of organizational structure is an inevitable product of the Fourth Industrial Revolution, and also a concrete manifestation of the implementation of future industrial management. As the industrialization deepens, the evolution of the industrial organization structure is accelerating, there is more refined and clearer division of labor than before, and the process of design projects also requires a better standardized operating system than before. Therefore, innovating the organizational structure becomes inevitable for enterprises. After all, the essence of organizational planning is to plan ahead.

CHAPTER 6

Beyond Industry 4.0

6.1 Technological Changes Trigger Employment Changes

As the core driver of the new industrial transformation, artificial intelligence is going to profoundly change how human beings fare and produce, and promote the overall leap of social productivity. Meanwhile, the impact of the widespread application of artificial intelligence on the labor market has aroused great concern in the society.

At present, the accelerated advancement of artificial intelligence on a global scale has attracted great attention from all the countries. Whether it is a simple mechanical action or a complex perception task, the power of artificial intelligence is remarkable. Moreover, as machine learning, big data, and computing capabilities advance, the efficiency and accuracy of artificial intelligence systems in performing tasks will also be enhanced.

It makes sense to be concerned—the breakthrough of artificial intelligence implies upcoming technical unemployment of all sorts of positions.

6.1.1 Ongoing "Machines Substituting Workers"

Till now, artificial intelligence has become the new engine of future technological revolution and industrial transformation. It drives and promotes the transformation and upgrading of traditional industries. And it has a wide range of applications, from industry and agriculture to financial education, from digital government to smart transportation, to judicial systems, medical care, and retail services. The impact of artificial intelligence on employment is becoming increasingly obvious.

Technically, because of the stronger computer capabilities, the greater data availability, and the improved machine learning and other algorithms, key technologies such as artificial intelligence are bound to evolve in the future. That machines replace human labor is not only happening today but also in the future. However, this has direct impact on the labor market, triggering new employment anxiety.

In fact, since the First Industrial Revolution, from mechanical looms to internal combustion engines, and to the first computer, the emergence of new technologies has always triggered the public panic of being replaced by machines. During the two Industrial Revolutions from 1820 to 1913, the share of American labor employed in the agricultural sector fell from 70% to 27.5%, and the number of today is lower than 2%.

Many developing countries are undergoing similar changes and even faster structural transformations. According to data from the International Labor Organization, the proportion of agricultural employment in China fell from 80.8% in 1970 to 28.3% in 2015.

During the rise of artificial intelligence in the Fourth Industrial Revolution, relevant research institutions in the U.S. released a report in December 2016 that in the next 10–20 years, the number of jobs replaced by artificial intelligence would rise from the current 9% to 47%.

A McKinsey Global Institute report shows that it is estimated that by 2055, automation and artificial intelligence will replace 49% of the world's paid positions. And India and China are expected to suffer the most. McKinsey Global Institute predicts that 51% of the work content in China is going to be automated, which is equivalent to 394 million man-hours.

From the perspective of the specific work content that artificial intelligence takes over, not only the vast majority of standardized and programmed labor can be done by robots, but even non-standardized labor in the field of artificial intelligence will be impacted.

As Marx pointed out, the means of labor appear as machines, and they immediately become a competitor with the workers themselves. Carl Benedikt Frey and Michael A. Osborne, both professors at the University of Oxford, predicted in their co-authored article that in the next two decades, about 47% of American employees will exhibit weak "resistance" to automation technologies.

In other words, the white-collar workers will also suffer a similar impact, similar to the blue-collar workers. In accounting, finance, education, and medical care, a considerable number of jobs will change their working modes with the advancement of artificial intelligence. Humans are responsible for the tasks that require high skills, creativity, and flexibility while robots take their advantages of speed, accuracy, and continuity to take care of repetitive tasks.

Obviously, although artificial intelligence cannot 100% replace the white-collar workers, it will inevitably cut down employment opportunities, so that the labor market is weak against automation technologies.

Meanwhile, in the face of the artificial intelligence boom, the preference for employment of highly skilled labor in a few cutting-edge innovation fields like high-end R&D continues, which has resulted in clear polarization in the employment of high-skilled labor and low-skilled labor: the demand for high-skilled labor soars, thereby intensifying the trend of removing low-skilled labor in general production.

MIT researchers used the labor market data of the U.S. from 1990 to 2007 to analyze the impact of the deployment of robots or automated equipment on employment and work. The results showed that in the American labor market, when there is a 1% increase in the proportion of robots in the total labor force, 1.8%–3.4% of jobs are lost. What's worse is that workers get paid less by an average of 2.5%–5%. The threat of technical unemployment is around the corner.

6.1.2   The Creation of Future Jobs

Certainly, this is not the first panic about automation in human history. Since the beginning of modern economic growth, people have periodically suffered intense panic over being replaced by machines. For centuries, this kind of concern has always turned out to be a false alarm. Despite the constant technological progress over the years, there will always be new jobs for people, which suffice to avoid large-scale permanent unemployment.

For example, there used to be specialized legal associates engaged in the retrieval of legal documents. However, since the introduction of software capable of analyzing and retrieving massive amounts of legal documents, the time cost has dropped sharply, thus greatly expanding demand. Therefore, the employment prospect of legal associates has not gotten worse but better (from 2000 to 2013, the number of employees for this position went up by 1.1% per year).

There is another example where the application of ATM machines once led to a large number of bank clerks being laid off. From 1988 to 2004, the number of clerks in each bank's branch in the U.S. dropped from 20 to 13 on average. However, the lowered cost to operate each branch allowed the banks to have enough funds to open more branches and better serve customers. As a result, the number of bank branches in American cities rose by 43% from 1988 to 2004, thus more people were hired as bank clerks in general.

History points out that technological innovation increases the productivity of workers, creates new products and markets, and generates new employment opportunities in the economy. And artificial intelligence may be no exception. From a long-term development perspective, it is going to create more jobs by reducing costs, and driving industrial expansion and structural upgrading.

Having analyzed the relationship between technological progress and employment in the UK since 1871, Deloitte concluded that technological progress creates jobs because it stimulates consumer demand for commodities by lowering production costs and prices. Consequently, the total social demand grows, the industrial scale expands, the industrial structure is upgraded, and more jobs are created.

From the perspective of the new employment space that artificial intelligence opens up, the first mode in which it changes the economy includes the creation of new products through new technologies and the development of new functions. Together, they stimulate new consumer demand in the market, thus directly creating a group of emerging industries and driving the linear growth of the smart industry.

Chinese Institute of Electronics believes that every robot produced can create at least four types of jobs, such as robot R&D, production, supporting services, quality management, and sales.

Nowadays, the mainstream drive for the advancement of artificial intelligence is big data. During the intelligent upgrading of traditional industries, lots of intelligent projects not only require sufficient data scientists and algorithm engineers but also more ordinary personnel for data processing such as data cleaning, data calibration, and data integration because manual operations are needed in this link.

In addition, artificial intelligence will also drive the linear growth of jobs in the smart industry chain. The intelligent development led by artificial intelligence will surely stimulate the development of related industrial chains and energize the upstream and downstream job markets.

In addition, with the enrichment of material products and the improvement of people's quality of life, the public demand for high-quality services, spiritual consumption, and high-end personalized services will continue to grow, which will create many new jobs in the service industry. McKinsey & Company believes that by 2030, the development of higher education and medical care will contribute to 50–80 million new jobs worldwide.

From the perspective of job skills, more simple repetitive tasks will be taken care of by artificial intelligence, and there will be plenty of jobs that require higher skills. This also means that even though artificial intelligence is driving industrial expansion and structural upgrading to create more jobs, in the short term, its impact on employment in the low- and medium-skilled labor market remains severe.

6.1.3   Response to the Challenges

The advancement of artificial intelligence has profoundly changed not only one or a few industries, but the production modes and consumption patterns of the entire economy and society. It will further cast a huge impact on employment.

Certainly, based on the multi-level and phased nature of the advancement of artificial intelligence, its substitution of human labor will take place gradually. However, addressing and coordinating its short-term and long-term impacts on employment is the key to the current and future response to machines' substitution of workers.

First, it is necessary to actively respond to the short-term or partial challenges to employment posed by the application of new artificial intelligence technology, such as formulating targeted measures to buffer its negative impact on employment, seizing the new industrial development opportunities that artificial intelligence creates, expanding the emerging industries of artificial intelligence, creating new jobs in related fields with the help of artificial intelligence, and giving full play to the positive role of artificial intelligence in employment.

To deal with the social problems of artificial intelligence, the market must be creative. Only the right incentive mechanisms and the right talents can hedge the huge impact of artificial intelligence on the job market. Since China's reform and opening up, the most outstanding outcome is the rise of thousands of entrepreneurs, who have stimulated economic growth. On top of that, the government was able to promote transportation construction and further propel the development of enterprises.

Second, great attention must be paid to the risks that new technologies may replace human beings on traditional jobs with the focus on the transfer and re-employment of mid-position staff. In fact, technologies do not entirely determine how much employment artificial intelligence eliminates, how much it makes, and what new jobs it creates. The system plays a decisive role there, too. In the fast-changing technological environment, it is the system that determines the ability and flexibility to help individuals and enterprises creatively create new jobs.

For example, if a person suffers unemployment, can his competence be transferred? How does the transfer work? These are the issues that the system needs to address. The government must support the establishment of non-governmental organizations that provide training for those who lose their jobs so that they get the necessary help to adapt to changes in job requirements.

Ultimately, jobs and the income they create are two different stories. Judging from the long-term impact of artificial intelligence on the labor market, close heed must be paid to its impact on the income gap of different groups, and the problems of middle-income groups' poor employment and shrinking incomes be solved.

Since the beginning of the 21st century, the labor market in some developed countries has presented a new polarization: there is a continuous increase in the employment proportion in both high-paying and low-paying occupations with a low degree of standardization and proceduralization while the employment proportion in middle-paying jobs with a higher degree of standardization and proceduralization is dropping. This is an employment income effect that differs significantly from previous technological progress. It puts the middle-income group in a more embarrassing employment situation than the low-income employees.

In this case, if the income distribution policy remains focused on the high-income and low-income groups as before but for the middle-income group, there will easily be uneven distribution to the new low-income group under artificial intelligence conditions, that is, middle-income groups earn as much as before or less due to technological progress.

At present, the impact of the wide application of artificial intelligence on the job market has triggered great social concerns. In response to artificial intelligence, it is not only necessary to govern the labor-capital relationship, but also to break away from the old industrialization logic of "the stronger get stronger," and get well prepared in advance to face the challenges with a broader vision, more dimensional methods, and more effective strategies.

6.2   Restructure the Global Value Chains

Modern economic globalization has experienced the transformation from Global Commodity Chains (GCC) to Global Value Chains (GVC).

GCC are a set of interrelated chained labor production processes that revolve around the final consumable commodity. Because of the fragmentation of the organization of complete commodity transactions and the deconstruction of the independent production factor systems of developed and developing countries, highly complex fragments of production and business activities can realize brand new cross-border connections on the basis of a large-scale and refined division of labor and reorganization. This has triggered the systematic reconstruction of the international production system after the 1990s.

Under such circumstances, value analysis is introduced into the GCC, which has become the GVC of today—a series of stages in which the production and sales of products and services increase the value of the product. At least two stages are completed in different countries. In other words, if a country, department, or company participates in (at least) one stage of GVC, it is part of GVC.

In the anti-industrial age, the international environment is getting more and more complicated. The outbreak of COVID-19 and the Sino-U.S. trade war together are catalyzing a new round of GVC adjustments.

6.2.1   From Expansion to Shrinkage

GVC is proposed to explain that the value-creating links of product R&D, design, production, marketing, and after-sales are subcontracted to different countries on a global scale, and that enterprises obtain correspondingly added value by participating in the different links of the product lifecycle. To some extent, GVC is the most important product of the economic integration of various countries in the past three decades.

Since the 1990s, as the General Agreement on Tariffs and Trade was upgraded to WTO, global production continued to grow, and GVC experienced great development both vertically and horizontally. As there is continuously more participation in the GVC, the length of the value chain is rapidly extending. Meanwhile, trade in intermediate goods have begun to surpass finished products, becoming a major part of international trade.

In this process, China has gradually evolved into a world factory and one of the centers of GVC and international trade. The unit labor cost (ULC) database of the Yicai Research Institute shows that in terms of participation in the GVC, China has outdone the traditional manufacturing powers such as the U.S., Germany, and Japan and become the world's strongest manufacturing power. At the same time, it has also become such a key link in the GVC that almost all industries are dependent on China to some extent.

McKinsey Global Institute once selected 20 basic industries and manufacturing industries, and analyzed the dependence of countries around the world on China's consumption, production, and imports and exports. It discovered that with the deep integration of Chinese manufacturing into the GVC, especially in the fields of electronics, machinery, and equipment manufacturing, China, in addition to playing the role of the "world factory" as a supplier in the GVC, is becoming a more and more important demand-side in the "world market."

One World Bank report shows that the GVC had the most rapid growth from 1990 to 2007. Technological advancement in transportation, information, and

207

communications, and the lower trade barriers convinced manufacturers to extend their production processes beyond national borders. Moreover, three regional value chain networks in Asia, Europe, and North America centered on China, Germany, and the U.S. respectively have come into being.

However, this GVC expansion halted after 2008. In 2008, the international financial crisis broke out. In that year, the share of GVC in global trade reached a peak of 52%. Then, it dropped instead. This happened in tandem with the slowdown in global trade growth.

According to the WTO statistics: since the 1990s, except 2001, the growth of global merchandise trade has remained at the level of 1.5 to 2 times the growth of global GDP. In the second decade of the 21st century, the situation changed.

In both 2012 and 2013, global merchandise trade growth was the same as the global GDP growth; in the following three years, the former was lower than the latter; there was a rebound in 2017 and 2018; in 2019, the global merchandise trade was in a state of stagnation due to the worsened persistent trade tensions between the U.S. and other countries, and it declined near the year end by 0.1% in general.

Today, the global trade growth rate is smaller than the sluggish global GDP growth rate. In the past boom period, the former was about twice the latter. The pandemic in 2020 hit the global trade hard. WTO predicted that the global trade would drop by 13% to 32% in 2020. The transformation of GVC is inevitable.

6.2.2 Multiple Factors Drive the GVC Transformation

There are many agents for the transformation of GVC.

First, after the global financial crisis, the economies of all countries have not recovered from the recession. Overcapacity has slowed down the growth of the world economy, and the slowdown in investment growth has been particularly outstanding. Meanwhile, the accumulative economic and social problems due to the rapid economic development in the past two decades after the financial crisis began to surface. In particular, the magnified and intensified structural contradictions about population and debt issues gave rise to the wave of global protectionism.

Second, major emerging market economies, such as China, have begun to extensively replace foreign intermediate products with domestic intermediate products, which enables the pure Chinese production activities to replace GVC production.

Also, in the past two decades, the Chinese manufacturing industry's labor productivity and unit labor costs (ULC, the labor cost required to produce a unit of added value; the higher the ULC, the less competitiveness) have both witnessed rapid rises. During the same period, the ULC of Japan and Germany, the third and fourth largest manufacturing countries in the world, continued to drop. This encourages the manufacturing industry to move out of China, thus affecting the changes in the GVC.

Third, developed countries have taken actions to attract the return of the manufacturing industry. Their governments have intervened more in the industrial transfer. Before the U.S. turned to "America First" trade protectionism, structural changes in the global economy had triggered a backflow of the manufacturing industry to developed countries to various degrees.

From 2011 to 2014, among the four countries of the U.S., Germany, France and Italy, the top four sub-sectors with the most active manufacturing backflow were chemical products, metal products, electrical products, and electronic products. Among them, the return of chemical products manufacturers was the most outstanding.

Trade protectionism measures such as higher tariffs and technological bans that the U.S. adopted have increased the costs of cross-border trade, of intermediate goods, and of the industrial chain, thus affecting the global production decision-making layout of multinational companies. This has sped up the return and relocation of some industrial chains, thus triggering the restructuring of the value chains, industrial chains, and supply chains around the world.

Last, the increasing use of labor substitution tools (such as robots) in manufacturing further reduces the need to allocate resources in the lowest-cost places in the world, thus causing a mismatch in the job market.

According to the research of Dalia Marin and Kemal Kilic, the GVC and the use of robots were positively correlated before the financial crisis. This means that in a better market environment, enterprises can lower the costs and expand production scales in two ways: deploying more robots and promoting the GVC.

After the financial crisis, the GVC and the use of robots present an obvious negative correlation. This suggests that when there is overcapacity, robots are more used as substitutes for the GVC. They also found that if COVID-19 increases the economic uncertainty by 300% and deceases the interest rates by 30%, the application rate of robots will soar by 76%, thus significantly shrinking the GVC.

Therefore, for a long time, countries have been actively responding to the challenges of the GVC. They aim to innovate or reconfigure the entire value chain in the context of global competition and restructuring.

6.2.3   GVC Reconstruction in the Pan-industrial Age

The pandemic in 2020 hit the GVC hard, thus accelerating its restructuring. For example, the pandemic and corresponding prevention and control measures have delayed or halted the production and transportation of intermediate products, so that enterprises are exposed to a greater risk of being unable to obtain key inputs. Many GVC participants with excellent productivity depend on timely delivery of inputs and lean inventory management, but these measures may make countries at the center of the GVC suffer the most in the pandemic.

In addition, the United Nations Conference on Trade and Development (UNCTAD) believes that the global outbreak of COVID-19 will affect global foreign direct investment (FDI), which in turn will impact the GVC. The world's top 5,000 multinational enterprises (MNEs) have lowered their profit forecasts for the year by 30% due to the pandemic, and this will go on.

The industries that have suffered the biggest blow are energy, basic metals, aerospace, and automotive. The profit forecasts of MNEs in developed economies have witnessed the most significant revision, down by 35%, much higher than the 20% in developing economies. The plummeting profits are expected to reduce global FDI by 30% to 40% while the foreign direct investment (FDI) of MNEs is the main impetus for the further deepening of the GVC.

But it should be noted that the pandemic, as a catalyst, has also sped up the development of the anti-Industrial Revolution characterized by digitization and informatization. As a new generation of information technologies such as the mobile Internet, the IoT, cloud computing, and big data advance, significant changes have taken place in the manufacturing and organizational modes of some industries, especially the high-tech industry. This has promoted the global decomposition, integration, and innovation of the GVC, greatly deformed the "smiling curve" of the international industrial division of labor, and changed added value of each link accordingly.

New elements have been added to the reshaping of the GVC in the post-pandemic era, extending it to a higher level of new economies and itself. And the high-tech industry represented by artificial intelligence, 5G, smart logistics, and online payment has begun to expand, exercising greater international influence.

The GVC reconstruction is an international division of labor that both upgrades and governs the value chain. In fact, with the profound changes in the international situation, this reconstruction is bound to happen. And the rapid technological advancement catalyzed by COVID-19 in 2020 has added technological elements to it.

If in the early discussion of "global competition and restructuring," "reconstruction" was considered the value chain participants' innovation in their value chain activities or the reconfiguration of the entire value chain, the current reconstruction of the GVC is also related to the vertical and horizontal "extension" and "shrinkage" of the division of labor in the value chain, as well as to the displacement of network nodes. It is also under the influence from the changes in the Industrial Revolution, technological progress, and global economic and trade rules.

National governments have arrived at a common understanding to seize the opportunity of global supply chain adjustment at the present stage. If the Chinese government wants to, in this wave of global value chain reconstruction, claim the dual commanding heights of economy and technology, it must select a suitable reconstruction path and promptly make a strategic design and policy for the chain change and path selection.

First, a country has to figure out its true position in the GVC. Then, it has to choose a suitable reconstruction path. There are three paths for a country to participate in the GVC reconstruction: actively embedding in the GVC, passively connecting to the national value chain, and leading the construction of regional value chains. In addition, the path selection varies due to the difference in the international competitiveness of a country's industries. The third step is to upgrade the technical level and improve the domestic supply chain network. The pandemic, regardless of its disruption of the GVC, has also created new opportunities, which will lay a new track for upgrading competition.

6.3   The New World Order

Technological progress and industrial transformation are important sources of human well-being.

Industrial production drives the rapid production and spread of scientific and technological knowledge, and promotes the modernization of people; the economies of scale and the economy of scope of the industrial production gather production factors and accelerate the urbanization of the human society; informatization greatly prevents the spatial distance from hindering communication, and strongly promotes the global division of labor in industrial production. In a sense, today's industrialized society and modern human life that is centered on urbanization are the concentrated results of every past Industrial Revolution.

The emerging new round of Industrial Revolution, characterized by the realization of intelligent interconnections between people, machines, and resources, is blurring the boundaries between the physical world and the digital world, and between manufacturing and services. It offers vast space for more efficient and eco-friendly economic growth driven by modern technologies.

Similar to the previous Industrial Revolutions, this round will also build a strong growth momentum for the global economy, profoundly change the economic structure and development mode of the countries around the world, provide new solutions to the dilemmas and problems in the human economic society, promote its leap, reshape the competition patterns among countries, and create a new world order.

6.3.1   Reshaping the International Competition Pattern

The Pan-industrial Revolution is becoming an important impetus for the world to rebuild the growth momentum and enhance human well-being. However, although the benefit from the Pan-industrial Revolution is global, there is an uneven distribution of value created by new technologies and new industries among countries.

Developed industrial countries expect to consolidate or even further strengthen their dominant position in the global economy by accelerating technological break-throughs and leading industrial development; developing countries that have already possessed a certain industrial foundation and technological capabilities also hope to seize the opportunities of the new Industrial Revolution to catch up and surpass them by exploring unique technological paths and business models. Therefore, competitions, and catching up and surpassing are inevitable and important themes of the Pan-industrial Revolution. There will be competition and selection in its advancing process, which will inevitably reshape the competition pattern among countries and create a new world order.

Judging from economic history, each round of Industrial Revolution undergoes two stages: an introductory period and an expansion period. In the introductory period, the innovation of new general-purpose technologies and enabling technologies is basically based on the accumulation and development of basic research. It exhibits obvious science-driven characteristics. Meanwhile, since neither the paradigm nor the path of new technologies are clear, different types of innovation entities, especially start-ups, driven by the huge potential interests, usually actively explore diversified technological routes and business models. When the general-purpose technologies and enabling technologies and matching business models mature, the application of these new technologies begins to encourage the emergence of new industries and accelerate their spread to other parts of the national economy. This is when the Industrial Revolution enters the second stage, the expansion period.

Both the introductory and expansion periods of the Industrial Revolution are precisely manifested in the incentive competition of countries or enterprises on the two levels of technology and business. During the introductory period, multiple technological routes compete with each other. Due to the uncertainty of the technological routes themselves and the high R&D investment required for them, no country can dominate all technological routes. Although some countries and enterprises have first-mover advantages in cutting-edge technologies and basic research, it remains uncertain whether they will eventually become the developers of the guiding technologies. In addition, since the development of information technology runs on a short cycle, if the last-mover countries can engage in high-intensity technological learning, it is highly likely for them to catch up and surpass the leading group technologically.

When the Industrial Revolution enters the expansion period, that is, both general-purpose technologies and enabling technologies are mature enough for wide commercial application, technological leaders may fail to adapt to the requirements of guiding technologies in time due to their national systems and strategies, thus losing the opportunity to transform from leading the technologies into leading an industry. The guiding technologies and leading business models are formed through repeated iterative market selections in the continuous feedback of the technologies and the market. The technological leaders may suffer defeat during their commercialization while the technological followers may use their market advantage or infrastructure advantage to laugh the last laugh in the market competition.

The massive economic dividend that the new Industrial Revolution may yield and its profound impact on the industrial competition patterns among countries have encouraged every one of them to be an active part of it. And the complexity of the technical and economic processes of the new Industrial Revolution puts infinite suspense into the competition results. Their competition and catching up and surpassing during the new Industrial Revolution will eventually be presented as the dynamic changes in their competitiveness and interest patterns.

Based on the experience of prior Industrial Revolutions, during the introductory period, the main source countries of general-purpose technologies and enabling technologies were the first to propel the transformation of basic scientific research achievement into technological applications. They explored different technological routes in an attempt to control the guiding technologies. In this process, their scientific research and technical merits have made each other stronger, thus enabling themselves to be the scientific and technological high ground of this new round of Industrial Revolution.

As the new Industrial Revolution leaps from the introductory period to the expansion period, guiding technologies have taken shape, and corresponding engineering and industrialization have become an international competition focus. At this time, countries with greater engineering capabilities and more creative business models make stronger competitors. Since the technological innovation of the new Industrial Revolution takes place in only a few countries, the technical merits between countries will suffer polarization, but the technological capabilities at this point have not yet been completely transformed into a country's industrial competitiveness and economic welfare.

Today, the new Industrial Revolution is in transition from the introductory period to the expansion period. The U.S., China, Japan, and Germany are the main promoters of the maturity and application of guiding technologies. As it enters the expansion period, general-purpose technologies and enabling technologies begin to spread widely and be applied. Those countries that have taken the lead in promoting the spread and application of guiding technologies in leading and induced industries enjoy improvements in their technological capabilities, production efficiency, economic growth, employment levels, and overall national strength, thus becoming the biggest beneficiaries of this new Industrial Revolution.

The new Industrial Revolution is a complex process of coordinated transformation of technological and economic paradigms. Technological advancement and industrial development are embedded in a country's national system and policies. Technological breakthroughs and industrial changes are bound to encounter institutional resistance as they alter the existing interest structures. Therefore, countries and regions that can adjust their systems and policies relatively faster to enable themselves to support the development of new labor skills, emerging technologies, new ventures, and leading industries more effectively, so as to better match the technical and economic requirements of the new Industrial Revolution, can become the main beneficiaries of it.

In this competition of national systems and policies, developed countries are trying to make use of the new Industrial Revolution to further enhance their industrial competitive advantages, halve the trend of "industrial hollowing," and regain the competitive advantages of manufacturing. In recent years, these countries or regions have issued mid-to-long-term development strategies for intelligentized, networked, and digitized manufacturing technologies, such as the Advanced Manufacturing Strategy in the U.S., the Industry 4.0 in Germany, and the New Industrial France in France, the Digitalized European Industry strategy in the EU, the Industrial Connectivity 4.0 in Spain, the New Robotics Strategy in Japan, the Manufacturing Innovation 3.0 in South Korea, and Italian Manufacturing in Italy. All of them reflect the grand ambitions of the developed industrialized countries to further strengthen their technological and industrial competitiveness.

In the meantime, developing countries represented by China are rapidly building a relatively complete industrial system and innovation system by undertaking industrial transfers and independent innovations. The broad participation of developing countries with a certain industrial foundation, including China, in the breakthrough and application of high-tech is the biggest distinction of the new Industrial Revolution from prior Industrial Revolutions.

In the new Industrial Revolution, not only were the late-moving countries given the chance to run alongside the developed countries in emerging industries, but also due to the integration of traditional technologies and traditional industries with new technologies, they opened the window to catch up and surpass the leading countries in mature industries by taking advantage of their unique market and resource advantages. The perfect example is that in the 1970s, when the automotive technical routes changed from low cost and power enhancement to diversification, energy saving, and environmental protection, Japanese enterprises surpassed both German and American automobile industries through flexible production and lean manufacturing. At present, the Chinese manufacturing power strategy, Russia's National Technology Plan, Argentina's National Production Plan, and India's India Manufacturing Strategy all voice the strong demands of developing countries to participate in the new Industrial Revolution, which will further intensify the international scientific and technological competition.

6.3.2 The New Direction of Global Governance

Marx believed that social relations are closely connected with productive forces. When there are new productive forces, people change their production modes. When that happens, meaning the ways to secure their lives change, people change all their social relations. The hand mill gave rise to a society led by the feudal lord, and the steam mill to a society led by industrial capitalists. The changes in the production mode determine the social changes and fundamentally the nature of the society. Therefore, the ultimate cause of all social changes and political changes is not to be found in people's minds, in their better understanding of the eternal truth and justice, but in changes in the production mode and exchange.

The rise of the Pan-industrial Revolution triggered changes in both the production mode and the exchange mode. The large-scale assembly line production mode has been transformed into the self-production mode of digital manufacturing. The new productivity platform enables intelligence, customization, and cooperation. This prompts revolutionary changes in the manufacturing industry, creates lots of new industrial clusters and economic growth points, expands the scope of strategic emerging industries, and pushes mankind to enter the stage of global socialized mass production with the help of the Internet.

This stage is characterized by global division of labor and cooperation, intelligent management, ecological harmony, and sustainable development. This upgraded socialized mass production, while fundamentally improving the low value-added situation of the processing and manufacturing links in the entire industrial value chain, will greatly enhance world productivity, accelerate the global economic development, promote the development of international economy and trades, change the global monetary and financial relations as well as the international flow of production factors, drive the international economic integration, and trigger changes in the global economic structures and strategies.

Obviously, the world of today is flat and connected. Although there is still international conflict, which is even constantly escalating, and local wars break out, it is an indisputable fact that extensive connections are built in the course of opening up. The integration of these vast connections in the course of changes will inevitably lead to a grand co-governance of the world.

The tragedy of the global financial crisis in 2008 has long taught the world a profound lesson: that the time when only a few major powers controlled the world is ending, the global balance of power is changing in the direction of benefiting the vast number of developing countries, and global governance is shifting from "West dominating the world" to "East and West co-governance." With the advancement of world multipolarization, economic globalization goes faster. Under the theme of peace and development, the international system will be flattened. The arrival of pan-industry not only reshapes the international competition patterns, but also allows countries to build a community of interests, of destiny. In this process, new industrial technologies will also give new solutions to some global issues.

While going through multi-polarization, economic globalization, cultural diversification, and social informatization, the world of today is exposed to non-stopping unconventional security issues, such as climate change, cyber-attacks, environmental pollution, and transnational crimes, that challenge human survival and the international order. Every earthling is in this community of common future. The interests and destiny of all countries are closely linked like a domino effect. No country can detach itself from the connected world or conquer it. The international community is becoming a community of a common future. Global governance issues directly puts the wisdom of mankind to the test, and the new Industrial Revolution has new possible solutions to them.

For example, the development and popularization of green energy have provided more effective solutions to environmental problems arising from population growth and industrialization: the development of auto-pilot and smart transportation will provide new technical routes for solving increasingly severe urban traffic problems; the development of cross-border e-commerce and other emerging business models brought about by digital technologies, as well as the greater convenience of service trade will effectively fuel the growth of global trade. *The World Trade Report 2018* from the WTO predicted that global trade would experience a 1.8%–2% increase year by year before 2030; the new Industrial Revolution would lead to a tearing down and re-assembly of the global value chains, supply chains, and industrial chains, thus further deepening the labor divisions and the improvement of transaction efficiency, and accelerating the recovery of the global economy.

The Pan-industrial Revolution has linked the interests and fate of all countries closer and tighter. Only by coordinating competition policies and social policies under the framework of win-win cooperation, and jointly solving the economic and social problems such as monopolies, zero employment growth, and social ethical problems that may come with the new technologies, can countries more effectively deal with the challenges that the new Industrial Revolution poses, and guide it in a direction conducive to solving major global problems and promoting inclusive global development.

6.4 The Stable and Promising Future of the Industry

Although the Pan-industrial Revolution promises an unprecedented technological future, we are in downward times. Political, economic, and cultural standards are aligned at the lowest point, which have become the new standards. With the abandonment of honor, the disgrace of dignity, and the pride in shamelessness, under the one-sided and one-size-fits-all mindset of national utilitarianism and pragmatism, the capital pursuit of profit is rampant. Therefore, in such a time, there is an urgent need for humanistic thoughts to guide people's spiritual lives.

6.4.1 Industrial Civilization Urgently Needs Correction by the Humanities

With the continuous technological advancement, people have moved from the age of steam to the age of electricity, to the age of atoms, and at last to the information age. However, while technologies enrich the material world, they also expose people to more and more crises of cognition, of survival, and of belief.

Since the 21st century, because of rapid advancement in science and technology, the structure and spiritual outlook of human society are constantly changing. Communication technology, the Internet, big data, cloud computing, blockchains, artificial intelligence, genetic engineering, and virtual technology have resulted in the integration of information and entities and a data-driven economy. The intelligent networking of the entire society is profoundly revolutionizing production mode, lifestyle, thinking patterns, and governance.

However, although emerging technologies opened the Fourth Industrial Revolution, they also trapped some people in (rational) conceit again. In real life, rationality degenerates into algorithms and calculations, and calculations deteriorate into plotting. While the new technological revolution and Industrial Revolution vigorously drive social development, they also expose people to environmental, ecological, and ethical risks, as well as personal spiritual loss, absence of belief, and the crisis of meaning. All of these are in urgent need for value reengineering through the spiritual guidance of humanistic thoughts.

Global warming is an undeniable fact. In 2020, a team of 93 scientists published a paleoclimate data record spanning the past 12,000 years. It contains 1,319 data records that are from samples of lake sediments, marine sediments, peat, cave sediments, corals, and glacial ice cores collected from 679 locations around the world.

With this data, researchers were able to map the changes in the air temperature on the earth's surface over the past 12,000 years, and compare them with the century average temperature from 1800 to 1900 to track possible changes that the Industrial Revolution caused. As expected, at the beginning of the period, the temperature was much lower than the baseline in the 19th century. But over the next few millennia, it rose steadily and eventually crossed the baseline.

The temperature peaked 6,500 years ago, and thereafter, the earth has continued to slowly cool down. The cooling rate after the peak is subtle, only about 0.1°C per 1,000 years. However, since the mid-19th century, human activities have lifted the average temperature by as much as 1°C, a significant peak value in a relatively short period, higher than that 6,500 years ago.

Climate change has disrupted the balance between the sun's light/heat on the earth and its reflection, and the most direct consequence of imbalance is the exacerbated occurrence of climate disasters. Germanwatch, an NGO, released a report in 2020 that analyzes storms, floods, and scorching weather without including "slowly surfacing environmental changes" such as rising sea levels, warming sea water, and melting glaciers. The report shows that between 2000 and 2019, there were about 11,000 extreme meteorological disasters around the world.

The infectious disease threat that was once thought to have been controlled is back. Part of climate change has expanded the geographic range of ticks and tick-borne pathogens. Due to the lack of global governance, policies, and international cooperation to slow down climate change and maintain a more balanced relationship between man and nature, there is a greater threat of the spread of tick-borne diseases and other infectious diseases. In addition, since the mid-20th century, the Arctic surface temperature has continued to rise, and the rate of increase is nearly double that of the global average. Rising temperatures have changed sea ice, snow cover, and permafrost, affecting the lives of about seven million people. For example, if mercury and other persistent environmental pollutants and infectious sources in the permafrost are thawed, these substances will be released and pose health risks.

The abuse of antibiotics together with natural evolution has created increasingly dangerous microorganisms. According to the U.S. Centers for Disease Control and Prevention, there are more than 2.8 million cases of antibiotic resistance each year in America alone, and more than 35,000 American die because of it. In India, neonatal infections caused by antibiotic resistance kill nearly 60,000 newborn babies every year. The UN fears that by 2050, the annual death toll from drug-resistant infection will be ten million worldwide.

Antibiotic resistance not only gravely harms human health, but also significant burdens and damages the economy. The U.S. medical system alone has to pay US$20 billion every year to solve the drug resistance problem. British economist Jim O'Neill predicted that by 2050, global antibiotic resistance would cause a cumulative economic loss of US$100 trillion. In addition, reports from the World Bank and the UN Food and Agriculture Organization also pointed out that should the problem of antibiotic resistance remain unresolved in 2050, the global annual GDP will fall by 1.1% to 3.8%, which is as horrible as the financial crisis in 2008.

It has a cost that computerization is constantly revolutionizing new areas of work and recreation, including the more unemployment, the wider digital divide, the collapse of traditional community formation and maintenance, and the inability of network interaction to completely replace traditional communities (for now, and perhaps never). The social structure is biased towards globalization, and regional cultures blend in all sorts of forms. Artificial intelligence is developed, and powerful systems rule everyone's life through various means.

In today's highly developed mobile Internet, social media with ubiquitous tags sketches the bizarre world based on algorithms; male and female anchors with collapsed character settings use technology to create an illusion of an eroded reality; all information, knowledge and opinions are within reach, and the massive amount of information, like randomly coded symbols, create a data ruin; all highlight the fact that in this era we are divided from others.

The anxiety that the construction of the legal and aesthetic systems falls behind the rapid development of science and technology terrifies the people in them. Whether it is the thousand-year-old traditional Chinese benevolence, righteousness, etiquette, or the democratic system under Western liberalism, they have been torn up with the advent of the Internet. Therefore, at the strategic pass of the age of machines, people, facing this enormous void, have been confused and hesitated about the unknown future in this age.

The ever-developing science and technology, while rewriting the world and people's mode of life and inherent values, have taken mankind to an unprecedented no-man's land. People truly discover and acknowledge that the religions, humanistic thoughts and values that used to guide can no longer sustain the changes brought about by technologies to some extent. In other words, it is the mankind's current thinking and interpretation of religions and humanities that cannot guide mankind to cope with the current and future no-man's land.

At the same time, the growing prosperity of pragmatism in modern commercial society has made the humanistic spirit dwindle. In the knowledge-driven economy, the added value of knowledge attached to the goods will lead to their obvious appreciation. Therefore, in a knowledge society, knowledge can only survive in application. Moreover, since World War II, natural sciences have been leading social development through the information revolution and the new technological revolution, while scientific and technological civilization has dominated the contemporary world. Compared with the natural sciences, the humanities appear abandoned.

As peace and development become the themes of the times, the humanities and social sciences fail to directly generate economic value in a short period of time and become an important force for promoting global economic growth and technological progress. In the meantime, because the humanities indulge in the past and fail to respond sharply to the modern society, they can no longer give more theoretical guidance and help in solving the complex problems under the new world situation.

From the perspective of specific higher education practices, the humanities and social sciences seem to be gradually marginalized: the University of Amsterdam released a school planning outline called Profile 2016 in 2014, which proposed to cut funding and remove some linguistic studies. Meanwhile, the remaining majors of the Faculty of Humanities, including philosophy, history, and Dutch literature, would be merged into the "humanities degree," and the construction focus of the school would lean towards career-oriented majors. U.S. funding for the humanities also dropped from US$400 million in 1979 (recorded in 2016 US$) to US$150 million in 2016 (recorded in 2016 US$).

These are the times when crises and new machines coexist. The development of industrial civilization urgently requires the guidance and correction of humanities.

6.4.2 From Integration to Renewal

The integration between technologies and culture and their arrival at a new height are in fact the organic combination of cultural content, ideas, and forms with technological spirit, approaches, and theories. It changes product value and quality and creates new content, form, function and service. This is a process of innovation and will renew the social order.

Robert King Merton, a representative of American structural functionalism, once placed technology in the visual thread of social change and discussed how it and culture affect the society. As Merton suggests, it is precisely the lack of the conceptual framework required by the social and cultural structure of technology itself that badly hinders technological development. This is because however the surrounding culture affects the development of scientific knowledge, or however science and technology ultimately affect the society, all these impacts come from the changing system and organizational structure of technology itself.

Although the humanities fail to replace the government and the public in formulating public policies and to deprive the public and the elected officials of the decision-making power, they can provide the public with information so that the public can make decisions wisely on the basis of correct information. This is also the basis for the survival of the future industrial world.

In the early 1970s, U.S. President Nixon initiated the development of supersonic transport. Back then, he wanted to make a technical achievement as outstanding as President Kennedy's Apollo Project, hence the supersonic transport plan, which was a government-funded cooperation with Boeing to develop supersonic aircrafts for civil use or bomber.

As a member of the Science Advisory Board, Melvin Calvin led a government advisory team to investigate this issue. They informed the president that considering factors such as economic benefits and environmental impacts (huge acoustic shocks and extremely high-altitude air pollution), the supersonic transport would do more harm than good. During the dispute, Calvin testified publicly in Congress against the project, so that it did not pass Congress.

As Calvin and other scientists pointed out in Congress, technological decision-making should not only stop at the technical rational level in the narrow sense, but extend rational and objective thinking to the technical level, and to the broader social, economic, and political levels for scrutiny. In other words, ethics, sociology, and even history and philosophical research must be able to keep up with the times in terms of social decision-making. If the advancement of science and technology is separated from the humanities, there will be inevitable damage to the entire human race.

In the social context of technology and culture, technology is a social system that is slowly taking shape and changing. Obviously, it not only benefits the entire society economically, but also performs cultural transformation and social order renewal, which are visual results from the beginning of the Industrial Revolution. This means that the Pan-industrial Revolution represented by information technologies such as artificial intelligence and big data will also reshape a new and ready-to-go social order, where there must be the social value concept of new and old values and their successful combination.

Inspired by the triple helix of genes, the theory of innovation triple helix, which suggests that to support the innovation system, there must be a helix connection mode, creatively proposes a new paradigm of innovation. This entangled helix is composed of three forces: an administrative chain composed of local or regional governments and subordinate bodies; a production chain composed of vertically and horizontally connected companies; and a technological science chain composed of research and academic institutions. In addition to performing the tasks of knowledge creation, wealth production, and policy coordination, the three departments also derive new functions through interaction, and ultimately innovate based on knowledge reproduction.

Obviously, science and technology, culture and social order are geared to the innovation triple helix theory, based on which, they develop. Among them, social order is the goal of cultural pursuit and technological exploration, but its explanatory significance is neither in culture, nor in technology. Social order is an independent and substantial helix.

In the triple helix of culture, technology, and social order, the three overlap and fuel innovation, forming "reciprocal causation." As the core attractions of the system formation, they enable the innovative triple helix to hold. On the other hand, the establishment of the social order, the dissemination of "dominant" culture, and technological progress promote the spiraling-up of the innovation helix so that they become the internal driving force for the innovation triple to hold.

On the one hand, the economic attributes of technological creation determine that technology must have a close connection with the external market. Meanwhile, the uncertainty of the external market reacts back to culture, technology, and social order. Therefore, culture is required to guide the path of truly promoting the healthy social development through the mutual promotion of culture, technology and social order.

On the other hand, the main driving forces that culture, technology, and social order form come from the respective needs of those three. The innovation of cultural excitation requires the support of science and technology and the guidance of values; the social order reflects the new ideas of the times, and requires the expression of cultural content and the support of science and technology; and the milestone leap of science and technology is inseparable from the overall cultivation of the social and cultural atmosphere and the guidance of the new social order.

On the surface, the process of human society is driven by one scientific and technological invention after another; in the middle, it is technological innovation that innovates culture, introducing new lifestyles to the human society; deeply, it is the integration and innovation of technology and culture that have created new human groups and bred a new social value system.

Obviously, the development of human civilization is by no means the growth of external materials, but the establishment of an inherent spirit, and the upgrade from individual self-discipline to the overall social order. When science and technology play a more important role in the industrial society, transitioning from a technical force to a commercial and economic power, how to effectively integrate technology and culture and turn them into an important support for social development and civilization progress will be an inevitable challenge for the times.

With eyes on the future, as the industrial society is restructured and the thinking pattern is reset, the world will change accordingly and present brand new meanings.

227

References

Chen, Liuqin. "A Review of Research on Global Value Chain Theory." *Journal of Chongqing Technology and Business University (Social Science Edition)* 26, no. 6 (2009): 55–65.

China Securities Co., Ltd. *In-depth Research Report on 3D Printing Industry.*

CICC. *Research Report on Energy Industry: Technological Breakthroughs Revolutionize Energy.*

Debon Securities Co., Ltd. *Research Report on the Industrial Robot Industry: The Boom Cycle Starts, Pioneering Intelligent Manufacturing.*

Fu, Jianzhong. "Development Status and Trend of Intelligent Manufacturing Equipment." *Journal of Mechanical & Electrical Engineering* 31, no. 8 (2014): 959–962.

He, Zhe, Sun Linyan, and Zhu Chunyan. "The Concept, Problems and Prospects of Service-oriented Manufacturing." *Studies in Science of Science* 28, no. 1 (2010): 53–60.

Huang, Qunhui, and He Jun. "'The Third Industrial Revolution' and the Adjustment of China's Economic Development Strategy—From the Perspective of Technical Economy Paradigm Changes." *China's Industrial Economics*, no. 1 (2013): 5–18.

Ji, Chengjun, and Chen Di. "Research on the Path Design of Furthering 'Made in China 2025' —Based on the Enlightenment of German Industry 4.0 and the U.S. Industrial Internet." *Contemporary Economic Management* 38, no. 2 (2016): 50–55.

Li, Jinhua. "Comparison between German 'Industry 4.0' and 'Made in China 2025' and Enlightenment Thereof." *Journal of China University of Geosciences* (Social Science Edition) 15, no. 5 (2015): 71–79.

Li, Xiaoli, Ma Jianxiong, Li Ping, Chen Qi, and Zhou Weimin. "3D Printing Technology and Application Trends." *Process Automation Instrumentation* 35, no. 1 (2014): 1–5.

Li, Yan. "Constraints and Promotion Strategies for the Development of Industrial Internet Platforms." *Reform*, no. 10 (2019): 35–44.

Liu, Fei, Cao Huajun, and He Naijun. "Research Status and Development Trend of Green Manufacturing." *China Mechanical Engineering*, no. Z1 (2000): 5, 114–119.

Liu, Jiejiao. "American Re-industrialization and Thinking Thereof." *Journal of the Central Party School of the Communist Party of China* 15, no. 2 (2011): 41–46.

Liu, Zhimin. "Lean Production and Its Status and Role in Advanced Manufacturing." *Modern Economic Information*, no. 15 (2016): 68, 70.

Lü, Tie. "Research on the Reform Path of China's Industrial Internet Industry—From the Perspective of Platform System Architecture." *People's Tribune·Academic Frontiers*, no. 13 (2020): 14–22.

Ping An Securities Co., Ltd. *3D Printing Industry and An Panoramic Analysis on Its Industrial Chain.*

Ren, Baoping, and Zhu Xiaomeng. "The Transformation of China's Economy from the Consumer Internet Age to the Industrial Internet Age." *Shanghai Journal of Economics*, no. 7 (2020): 15–22.

Research Group of the Institute of Industrial Economics of the Chinese Academy of Social Sciences, and Lü Tie. "The Third Industrial Revolution and China's Manufacturing Strategy." *Learning & Exploration*, no. 9 (2012): 93–98.

Shenwan Hongyuan Group Co., Ltd. *Research on the Industrial Robot Industry: The Inevitable Boom of Industrial Robots in China, and Domestic Manufacturers Fully Enjoy the Industry Dividends.*

Sun, Linyan, Li Gang, Jiang Zhibin, Zheng Li, and He Zhe. "Advanced Manufacturing Mode in the 21st Century—Service-oriented Manufacturing." *China Mechanical Engineering*, no. 19 (2007): 2307–2312.

Wang, Yanan. *Research on the Competitiveness of China's Manufacturing Industry from the Perspective of the Integration of Industrialization and Industrialization.* Beijing University of Posts and Telecommunications, 2011.

REFERENCES

Wang, Zhanxiang, Wang Qiushi, and Li Guomin. "Analysis on the De-industrialization and Re-industrialization of Developed Countries." *Modern Economic Research*, no. 10 (2010): 38–42.

Wei, Zhongjiang. "The International Political Effect of the Second Industrial Revolution." Insight, no. 19 (2016): 336, 338.

World Economic Forum. *The Fourth Industrial Revolution, the Light of Technological Innovation in Manufacturing.*

World Economic Forum. McKinsey Global Lighthouse Network. *The Latest Insights from the Forefront of the Fourth Industrial Revolution.*

Wu, Lei, and Zhan Hongbing. "International Energy Transition and China's Energy Revolution." *Journal of Yunnan University (Social Sciences Edition)* 17, no. 3 (2018): 116–127.

Xie, Fuzhan. "On the Accelerated Expansion of the New Industrial Revolution and the Direction of Global Governance Reform." *Economic Research* 54, no. 7 (2019): 4–13.

Xu, Yimin. *Research on the Influencing Factors of the Integration of Informatization and Industrialization.* Nanjing University, 2013.

Yan, Deli. "The Development of the Digital Economy Is Moving Towards a New Stage of the Industrial Internet." *iChina*, no. 6 (2020): 5–9.

Yang, Jing. "Prospects for the Development of China's Energy Technology." *Hubei Nongjihua*, no. 10 (2020): 50–51.

Ye, Xianming. "Marx's Industrial Civilization Theory and Modern Significance Thereof (Part 1)." *Studies in Marxism*, no. 2 (2004): 41–46.

Zhang, Shaojun, and Liu Zhibiao. "The Industrial Transfer of the Global Value Chain Model—Motivation, Influence, and Enlightenment to China's Industrial Upgrading and Regional Coordinated Development." *China Industrial Economics*, no. 11 (200): 5–15.

Zhang, Xiangyang, and Zhu Youwei. "Research on Industrial Upgrading from the Perspective of Global Value Chain." *Foreign Economics & Management*, no. 5 (2005): 21–27.

Zhao, Ruyu, Yan Guolai, and Guan Yuejia. "De-industrialization and Re-industrialization: the Experience and Lessons of Major European Countries." *Contemporary Economic Research*, no. 4 (2015): 53–59.

REFERENCES

Zhou, Jiajun, and Yao Xifan. "Advanced Manufacturing Technology and the New Industrial Revolution." *Computer Integrated Manufacturing Systems* 21, no. 8 (2015): 1963–1978.

Zhu, Shidong, Xu Ziqiang, Bai Zhenquan, Yin Chengxian, and Miao Jian. "Research Progress of Nanomaterials at Home and Abroad II—Application and Preparation Methods of Nanomaterials." *Heat treatment Technology and Equipment* 31, no. 4 (2010): 1–8.

Zou, Caineng, Pan Songqi, and Dang Liushuan. "On Energy Revolution and Science and Technology Mission." *Journal of Southwest Petroleum University (Natural Science Edition)* 41, no. 3 (2019): 1–12.

Index

ABOUT THE AUTHOR

KEVIN CHEN is a renowned science and technology writer and scholar. He was a visiting scholar at Columbia University, a postdoctoral scholar at Cambridge, and an invited course professor at Peking University. He has served as a special commentator and columnist for the *People's Daily*, CCTV, the China Business Network, SINA, NetEase, and many other media outlets. He has published monographs in numerous domains, including finance, science and technology, real estate, medical treatments, and industrial design. He currently lives in Hong Kong.